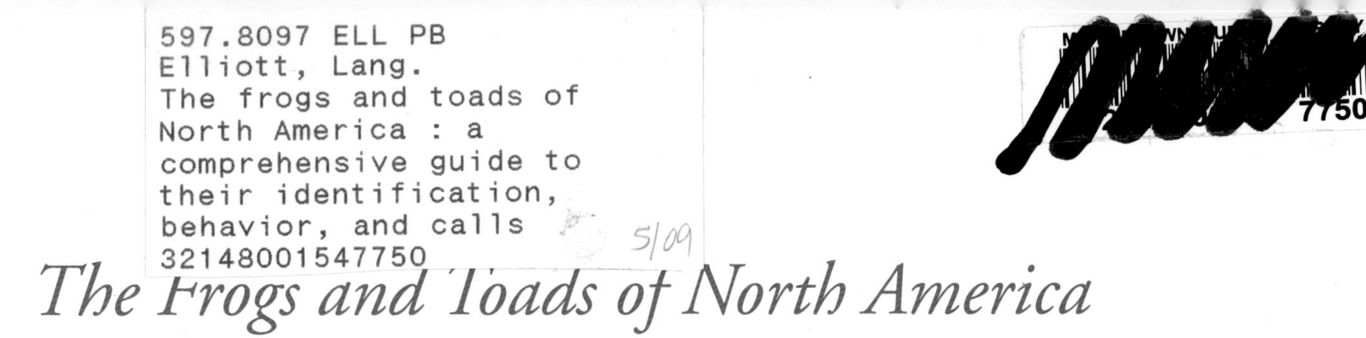

The Frogs and Toads of North America

A Comprehensive Guide to Their Identification, Behavior, and Calls

American Toad

The Frogs and Toads of North America

A Comprehensive Guide to Their Identification, Behavior, and Calls

Pig Frog

Lang Elliott, Carl Gerhardt, and Carlos Davidson

Houghton Mifflin Harcourt
Boston New York
2009

Squirrel Treefrog

www.hmhbooks.com

Library of Congress Cataloging-in-Publication Data

Elliott, Lang.
The frogs and toads of North America : a comprehensive guide to their
identification, behavior, and calls / Lang Elliott, Carl Gerhardt,
and Carlos Davidson.
p. cm.
Includes index.
ISBN-13: 978-0-618-66399-6
ISBN-10: 0-618-66399-1
1. Frogs—United States—Identification. 2. Toads—United States—
Identification. 3. Frogs—Canada—Identification. 4. Toads—Canada—
Identification. I. Gerhardt, H. Carl. II. Davidson, Carlos. III. Title.
QL668.E2E585 2009
597.8097—dc22
2008026090

Book design by Lang Elliott,
NatureSound Studio, P.O. Box 84,
Ithaca, New York 14851-0084

Cover Photo: Green Treefrog by Lang Elliott

Printed in China

C&C 10 9 8 7 6 5 4 3 2 1

Green Frog

Contents

Southern Cricket Frog

Foreword

About a hundred species of frogs and toads are found on the North American continent north of Mexico, providing a diversity of seasonal calls that are fascinating to most people but often difficult to sort out. Lang Elliott, Carl Gerhardt, and Carlos Davidson have addressed this situation in an exemplary fashion, with excellent recordings and exquisite photography accompanied by an informative and organized text, all bundled together in a book and compact disc that will provide hours of enjoyment for people who like to spend their time outdoors.

Much of that time will be spent listening on humid nights, when male frogs and toads are active and calling in the relentless search for mates that ensures the continued existence of their kind. The very best time to encounter these amphibians is shortly after a rainstorm, when the wind and noise have relented and the ground is covered with vernal pools. At that time, anurans come into full symphony and provide an enlightening view of their abundance. Many of us have had the experience of discovering the occasional toad sitting patiently beneath a streetlamp, as it waits to snack on a fallen insect, or have been startled by the leap of a sleek frog along the edges of a pond as it sought escape into deeper waters. But few of us can appreciate the full-blown cacophony of a mixed chorus of anurans on a spring night in eastern North America, the bleating song of Great Plains Narrowmouth Toads across the central grasslands of our continent, or the raucous din of spadefoots after a flash flood on our southwestern deserts. Folks who get to hear such noises are often stunned by the intensity of these little animals, but they shouldn't be—after all, it's about sex, and anything having to do with matters of the heart can be pretty compelling, even for a frog or toad.

In July 2003, Lang Elliott came to visit me and my wife, Suzanne, primarily so he could view and borrow some of Suzanne's amphibian images to use in this book. We were impressed by his intensity as he talked about frogs and toads and by his vision for a book about them. To be sure, Lang appreciated them, their beauty, and their unique position in the animal kingdom—wee vertebrates always conducting a balancing act with one foot on land and the other in water. But he was also keenly aware of their precarious place in the world, a world where water is now a precious commodity and where many folks are less and less willing to share it with the smaller creatures of the earth, or even to keep it clean for them. One evening, he and Suzanne and I mused about this, about the prediction of Rachel Carson in her profound book *Silent Spring,* about the bleak future for amphibians. It was not an uplifting discussion; we concluded that people still do not understand that we drink and share and swim in the same water as frogs and toads.

Maybe this book will change that. Maybe. But certainly it will help us to monitor the noisy little amphibians that share our water. It's important that frogs and toads stay noisy. The alternative would be unbearable—a silence that would mean the water was gone . . . or undrinkable.

Joseph T. Collins,
coauthor, *Peterson Field Guide to Reptiles and Amphibians of Eastern and Central North America*
Lawrence, Kansas, December 2008

Oak Toad

Introduction

Warm rains arrive, and there is an explosion of activity in the amphibian world; frogs and toads emerge from shelter and appear as if by magic in flooded pools, ponds, and streams. Individuals of one or more species join together in what appears to be a great celebration of sound—their yearly breeding effort, when males call excitedly to attract females in the age-old quest to reproduce.

Frogs and toads produce an impressive variety of sounds, all manner of croaks, peeps, trills, snores, barks, and chuckles. Choruses often pulsate with complex rhythms; neighboring males call back and forth in tight alternation and groups erupt after long periods of silence. While scientists interpret these calls in terms of their function—mate attraction, aggression, distress, and the like—the poet listens with a different ear, judging the emotional impacts of the sounds and the feelings evoked by the choruses.

The enchanting calls of frogs and toads emanating from wetlands in the dark of the night have a primal, timeless quality and evoke in many a sentiment expressed by Sigurd F. Olson in his book *The Singing Wilderness:* "This is a primeval chorus, the sort of wilderness music which reigned over the earth millions of years ago . . . one of the most ancient sounds of the earth, it is a continuation of music from the past, and, no matter where I listen to a bog at night, strange feelings stir within me."

In this book, we celebrate the lives and calls of more than one hundred species of frogs and toads found in North America. They are a unique and diverse group of organisms that many of us take for granted. But some species are in trouble, and others are likely to follow. Frogs and toads are indicators of environmental health. They are affected not only by habitat destruction and global climate change but also by chemical pollution and disease. A number of western species are undergoing severe declines and could be headed toward extinction, in part because of chytrid fungus, a disease that is having enormous impacts on amphibian populations throughout the world. Now is clearly the time for an increased awareness of our frogs and toads, coupled with closer monitoring of their populations and intensive scientific study of the causes of their declines.

Our frogs and toads are a natural treasure worth saving. They excite our imaginations; their sounds stir the music within our souls. They impress us at every turn, not only during the breeding season, when their calls enliven the night air, but also during our daytime walks along the shores of ponds, lakes, and woodland pools, when we share the experience of Basho and other poets of centuries past:

> Old dark sleepy pool
> Quick unexpected frog
> Goes plop! Watersplash.
>
> —Basho (seventeenth-century haiku poet)

Green Frog

Carpenter Frog

Classification

Frogs and toads are members of the order Anura, which is Greek for "tailless" (*an,* "without"; *oura,* "tail"). They are often referred to as anurans and there are nearly 5,300 species worldwide. In North America (north of Mexico), there are currently 97 native species and 4 introduced species of frogs and toads, for a total of 101 species. Scientific research into the taxonomy and classification of anurans continues, and the number of species is expected to grow, as new species are discovered with the help of modern genetic, biochemical, morphological, and behavioral analyses.

The frogs and toads of North America fall into nine family groups, which are described briefly below. More information about these can be found in the family introductions in the main body of the book. Refer to the photographs on the following two pages to see representative members of each group.

Hylidae: This is a large and diverse family, thirty species of which are found in North America (north of Mexico). The group includes the treefrogs, chorus frogs, cricket frogs, and several tropical species that just barely range into the United States.

Ranidae: Referred to as the true frogs, twenty-eight species live in North America. This family includes many of the familiar pond and lake frogs, including the Bullfrog, Green Frog, and various species of leopard frogs.

Bufonidae: Known as the true toads, twenty-two species are found in North America. This family includes the common hop toads, recognized by their chunky bodies, large wartlike glands on their skin, and terrestrial habits.

Scaphiopodidae: This small family is composed of seven species of North American spadefoots. They are a group of chunky, toadlike burrowing anurans with spade-like projections on their hind feet that aid in digging.

Microhylidae: This large family includes three species found in North America. Known as narrowmouth toads, they are small, burrowing amphibians with pointed snouts and narrow mouths.

Leptodactylidae: This is a large family of New World tropical frogs; only seven species range into the United States, mostly in Texas, Arizona, and Florida. Nearly all members of this group lay their eggs on the land; the larval stage is passed inside the egg.

Ascaphidae: Known as the tailed frogs because males have taillike copulatory organs, this family includes only two species, both of which are restricted to fast-flowing mountain streams in parts of the Northwest.

Rhinophrynidae: This family consists of a single species, the Mexican Burrowing Toad, which is found in the United States only in extreme southern Texas along the lower Rio Grande River.

Pipidae: Known as the tongueless frogs, this large and diverse family is represented in the United States by a single introduced species, the African Clawed Frog, which has become established in southern California and other locations.

Cane Toad

A Visual Guide to Family Groups

Class Amphibia
Order Anura—Frogs and Toads

North American species belong to nine family groups:

Hylidae—Treefrogs and Allies

Bufonidae—True Toads

Ranidae—True Frogs

Scaphiopodidae—
North American Spadefoots

Microhylidae — Microhylid Frogs and Toads

Leptodactylidae — Neotropical Frogs

Ascaphidae — Tailed Frogs

Rhinophrynidae — Burrowing Toads

Pipidae — Tongueless Frogs

Natural History

Frogs and toads (anurans) are amphibians—animals that normally live two lives, the first in water and the second on land. Eggs are laid in the water and hatch into aquatic larvae or tadpoles that are legless, breathe with the help of internal gills, and have prominent keeled tails to aid in swimming. After a variable period of growth, the tadpoles absorb their tails, sprout limbs, lose their gills, and transform into lung-bearing adults that may spend most of their time on land, returning to water only to breed.

While most frogs and toads adhere to this general life-cycle pattern, there are some exceptions. For instance, in species such as the Bullfrog and Pig Frog, adults remain more or less aquatic, wandering away from water only during rainy or humid nights. And there are terrestrial species, such as the various chirping frogs, that skip the aquatic tadpole stage altogether and lay eggs on land that hatch into miniature land-dwelling adults.

Frogs and toads as a group tolerate an amazing range of conditions. Southern Leopard Frogs are normally found in freshwater habitats but are able to survive in brackish marshes. Carpenter Frogs thrive in highly acidic bogs. New Mexico Spadefoots live in blazingly hot deserts where the lack of rainfall can postpone breeding for years. One frog, the Wood Frog, ranges northward all the way to the cold tundra of Alaska. Others show wide tolerances in both temperature and altitude; the Red-spotted Toad is found from below sea level in Death Valley to elevations exceeding 9,000 feet in the Rocky Mountains.

New Mexico Spadefoot

WATER LOSS AND ACTIVITY PATTERNS

Even though frogs and toads live in a wide variety of habitats, their daily and seasonal activity patterns are limited because they have permeable skin, which makes them prone to drying out—water evaporates from their bodies much more rapidly than in land animals whose skin is covered with scales, feathers, or hair. To reduce evaporative water loss, most species limit their activity to the night, when the relative humidity is higher and temperatures are cooler. During the day (and at night during dry periods), frogs and toads take shelter in moist areas under rocks, boards, and other objects, and by hiding in burrows; some of these burrows they make themselves, while others are inherited from or shared with other animals. If directly exposed to dry conditions, they may tuck in their legs and roll up or hug the ground, minimizing the surface area in contact with air. In extremely arid environments, numerous species estivate during the driest times of the year, resting in burrows and lowering their metabolic rates in order to survive the dry period.

One advantage of having permeable skin is that the skin can act as an important respiratory organ, directly

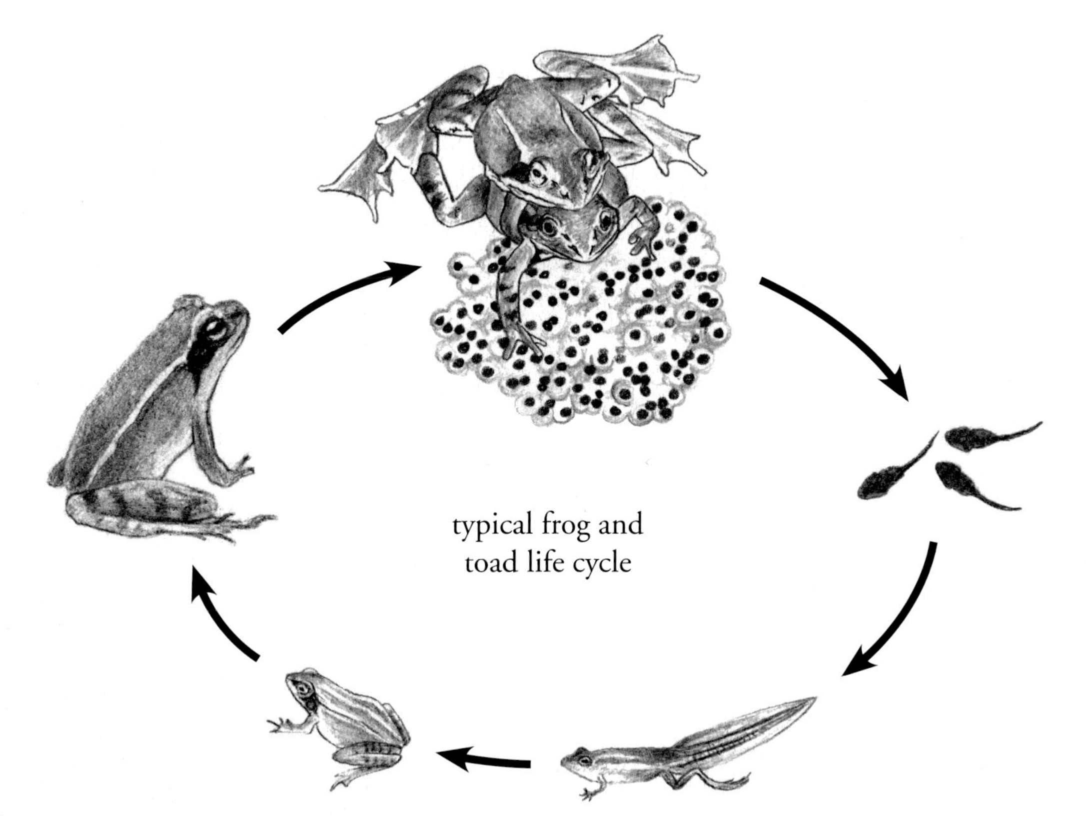

Drawings by Cindy Page

passing oxygen from the air to capillaries. This is especially helpful during inclement weather—respiring through the skin takes some load off the lungs, heart, and circulatory system and also helps frogs and toads survive long periods without eating.

THE CHALLENGE OF BEING COLD-BLOODED

Frogs and toads are cold-blooded and therefore are unable to regulate their body temperatures as warm-blooded animals can. When frogs and toads get cold, they become inactive. Consequently, activity in many species is limited to spring and summer, when temperatures are well above freezing. In cold climates, including at high elevations, frogs and toads are forced to hibernate for a substantial period each year, and they survive by burrowing deep into soil or mud, where frost cannot penetrate. Interestingly, Wood Frogs and several other species that live in areas with severe winters can actually tolerate being frozen—they are able to reduce the water content of their tissues and elevate the concentration of glucose or glycogen, which acts like a natural antifreeze and prevents damage to internal organs.

Although cold-blooded, individuals may regulate their body temperature to a limited degree by moving to warmer or cooler spots within water or by moving in or out of the sun. Adults of some species, such as the Canyon Treefrog, may raise their body temperature by basking directly in the sun during the day, but they typically do this near streams, pools, and other water sources and thus can easily replenish the water that evaporates from their bodies.

A positive consequence of being cold-blooded and having an overall low metabolic rate is that, unlike many warm-blooded animals, frogs and toads do not have to be constantly active and feeding. This allows them to live in extreme environments, although it limits their activity to times when conditions such as temperature and moisture are favorable and to seasons when food is abundant.

DIET

The two-part life of frogs and toads is reflected in their diet: tadpoles are vegetarian, and adults are carnivorous. Aquatic tadpoles subsist primarily on algae and other plant life, and they have specialized mouthparts that allow them to gather such food. But there are exceptions. In some species, such as desert spadefoots, crowded conditions may foster the development of cannibalistic tadpole morphs that prey on smaller tadpoles of the same species.

Adult frogs and toads subsist primarily on insects, but in general they will try to eat anything that moves and is smaller than themselves. Thus, larger species, such as the Bullfrog, eat not only insects but also small frogs, snakes, turtles, birds, and mammals. Their usual method of feeding is to sit and wait for moving prey rather than to search actively.

DEFENSE AGAINST PREDATORS

Frogs and toads have a wide variety of predators and defend themselves in various ways. Perhaps the most common tactic is to freeze and use their natural camouflage to avoid being seen. The ability of some species to change colors may help in this respect by improving their match with their surroundings. When pursued by a predator, a frog or toad, especially one belonging to a species with relatively long legs, can escape by leaping into a nearby body of water or by retreating into a burrow, crevice, or other protected spot. If the anuran is caught by a predator, noxious and sometimes toxic skin secretions may help it escape. Some species even produce loud distress calls; the sound may startle a predator or attract another predator, which may allow the frog or toad to escape as the predators interact. Still another tactic, one frequently adopted when a toad is caught by a snake, is to inflate as much as possible. This makes it difficult to swallow the toad and also exposes the toad's toxic parotoid glands.

Tadpoles avoid detection by being camouflaged and becoming still. Some, especially those living in places with predatory fish, are toxic. In several species, the tadpole's tail has a distinctive border that makes it more obvious than the body, leading a predator to attack the tail rather than the head. The Gray Treefrog employs a similar strategy—a high incidence of tadpole-eating insect larvae in a pool induces the development of broader and more colorful tails among the tadpoles.

AMPLEXUS AND FERTILIZATION

The most prevalent mode of reproduction in North American frogs and toads is external fertilization of eggs; the male is positioned on top of the female and clasps her with his forelimbs. This arrangement is called amplexus, and there are two basic forms. In inguinal amplexus, found in the tailed frogs, spadefoots, and the Mexican Burrowing Toad, the male clasps the female around her waist (inguinal amplexus is considered to be a primitive

A Barking Treefrog being stalked by a Pig Frog. Right after this photo was taken, the Pig Frog leaped at the treefrog and grabbed its leg, but the treefrog escaped.

A pair of Wood Frogs in amplexus at a communal egg-laying site

behavior). In axillary amplexus, found in all other species, the male clasps the female behind the front legs. In both cases, the male generally sheds sperm on the eggs as the female lays them.

Internal fertilization among frogs and toads is rare worldwide, and in North America only a few species accomplish this. The "tail" of the Rocky Mountain and Pacific Tailed Frogs is actually a copulatory organ that the male inserts into the female during mating. This mode of fertilization may be an adaptation to the swift-flowing streams in which these frogs live and breed. Amplexus in tailed frogs may last up to ninety hours, and the female stores sperm in her oviducts and uses it months later to fertilize eggs as she lays them. Internal fertilization is also typical of the Puerto Rican Coqui. In this species, the male sits on top of the female but does not clasp her, and the female loops her hind legs over those of the male, resulting in their two vents being pressed together and allowing sperm to enter the female. The pair often remain in this position for more than two hours before egg-laying begins.

EGG-LAYING AND HATCHING

Once mated, females of most North American frogs and toads lay their eggs in water, and the exact manner in which eggs are laid—such as in a large single mass, in scattered clumps, or in strings—is characteristic of each species. After hatching, the aquatic tadpoles may transform into small frogs quickly (in less than two weeks, for some spadefoots), after a few weeks or a month (for most frogs and toads), or after two years or more (for some true frogs and tailed frogs). Chirping frogs and other members of the genus *Eleutherodactylus* deviate from this pattern. They mate on land and lay their eggs in moist and protected spots. Their larval stage is passed within the eggs, and they hatch as tiny, fully formed frogs.

VOCAL COMMUNICATION

Choruses of calling frogs and toads are truly one of nature's most impressive events, especially in the southern states, where heavy rains may result in congregations of thousands of individuals representing nearly a dozen species. Although many frogs and toads produce harsh, mechanical-sounding calls, the musical notes of such species as Spring Peepers and Bird-voiced Treefrogs rival the beauty of some bird songs. While people who frequent nature enjoy the choruses and know that species can be recognized by their distinctive sounds, many do not realize that most frogs and toads have more than one kind of call. Furthermore, each call type has a different function and is accompanied by interesting behavior.

VOCAL REPERTOIRE

The most prominent call in a male's repertoire is his advertisement call. These loud, long-range sounds attract females that are ready to mate and lay eggs. In a typical species, this call also conveys a male's location and his readiness to defend his territory or calling space. The result is that males space themselves throughout the breeding site, each male defending at least a small area around himself.

If calling neighbors get too close, a fight may result. But fights can be avoided when a male switches to a distinctive aggressive call, which may result in the other male's peaceful retreat. Aggressive calls may be relatively

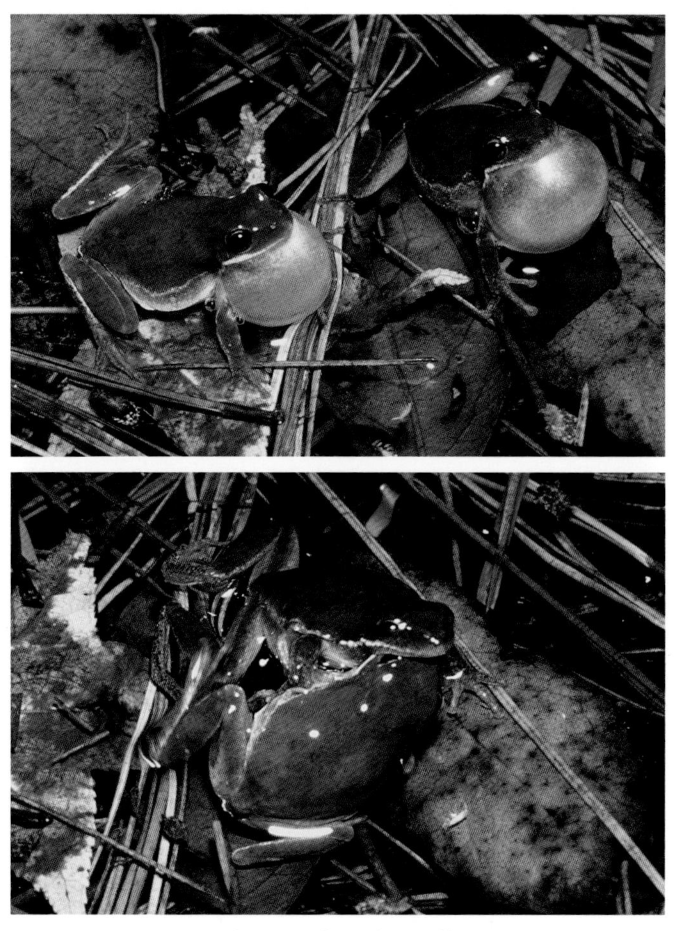

Two male Squirrel Treefrogs calling next to one another, and then grappling

simple modifications of advertisement calls, such as the *quarr-quarr-quarr* of the Green Treefrog, which sounds like a rapidly repeated hoarse version of its typical *quonk* call. Or they may be very different in quality: the *weep* notes of the Gray Treefrog sound totally different from that species' musical trills. Some species, including most true toads, do not produce aggressive calls at all.

When a male mounts another male or mounts an unreceptive female (one who has laid eggs or is not ready to lay), the mounted individual often responds with a release call, which may be accompanied by vibrations of the midsection. In some species, such as the Green Treefrog, release calls sound very much like aggressive calls, while in other species, such as the Bird-voiced Treefrog, they are distinctly different.

Courtship calls are often simple modifications of advertisement calls and may be produced by males that detect females moving toward them. For instance, Spring Peeper males respond to females by lengthening their advertisement calls, while Gray Treefrog males produce one or two calls that are as much as four times longer than normal. In Green and Squirrel Treefrogs, males simply increase the rate at which they produce advertisement calls. While many examples exist from other parts of the world, male courtship calls that are clearly different from advertisement calls are rare among species found in North America, the only known example being the soft courtship notes produced by males of the introduced Puerto Rican Coqui. Female frogs and toads may also produce sounds during mating. For instance, female Bullfrogs and Carpenter Frogs produce soft courtship notes just prior to amplexus. Because these calls occur only during a brief period and are not easy for humans to hear, it would not be surprising if females and males of other species are eventually found to have similar calls.

Distress calls are sometimes produced when a frog is seized by a predator (see Defense against Predators, page 18). These calls are typically loud screams that are made with the mouth open. Good examples are the wailing screams produced by Bullfrogs and Pig Frogs. There is little evidence that other frogs pay any attention to these calls; the current speculation is that the calls may startle the predator, possibly causing it to drop its prey, or may attract other predators, allowing the captured individual to escape.

MATING SYSTEMS

Male frogs and toads secure matings with females in a number of different ways. In the majority of species, males gather at a breeding site and space themselves so that their calls can be clearly detected and they can be found by females. When a female approaches a calling male, he usually stops calling and clasps her, although she may first initiate amplexus by nudging him. Once mounted, the female carries the male to an egg-laying site of her choice. Quite often, this site is well away from the place where amplexus occurred. Because calling and mate attraction typically do not take place where eggs are deposited, this kind of mating system has been likened to the lekking behavior of birds such as prairie-chickens, whose males gather and display in particular areas called leks. But in the anurans' case, females carry their partners to the spots where they lay their eggs; in prairie-chickens, females leave their mates at the lek and wander off to nest and lay eggs on their own.

A different kind of mating system may occur in species with prolonged breeding seasons, such as Bullfrogs and Green Frogs. In these species, a male uses advertisement and aggressive calls, as well as outright fighting, to defend a relatively large territory in which mating and

A male American Toad trying to mount another male, who responds with release calls

egg-laying occur. Such territories are usually defended for the entire breeding season. In the Bullfrog, males interact aggressively at the beginning of the season and often have prolonged wrestling matches that help them establish their individual territories. Then things settle down as males learn the advertisement calls of their neighbors and show reduced aggression toward one another. However, if an established male is removed and replaced by a new calling male, the neighbors instantly recognize the change and react with high levels of aggression. Neighbor recognition, which is also common in songbirds, saves time and energy and reduces the risk of predation when males are engaged in aggressive behavior.

In so-called explosive breeders, such as spadefoots and many true toads, most calling and mating takes place over the course of one or two nights, in temporary ponds, pools, and ditches after heavy rains. Males of these species typically do not constantly defend areas around themselves, nor do they have aggressive calls in their repertoires. In dense aggregations, they move about constantly in what is referred to as scramble competition, grabbing any other frogs that they detect, regardless of sex. Once a male has successfully mounted a gravid female, scramble competition may continue as other males pile on, all vying to get the most secure grip (in such cases, it is possible that the female's eggs will be fertilized by more than one male). Scramble competition is easily observed in such species as American Toads, where one is alerted to the numerous mismountings by squeaky release calls. It is also common to find two or more males grasping a single female (and one another) to create a "toad knot" (see photo to right).

In the breeding choruses of many species of toads, treefrogs, and some true frogs, there are breeding males that do not call. These so-called satellite males usually situate themselves near a calling male and then attempt to intercept females that are attracted to him. In some species, satellite males are noticeably smaller than calling males and may adopt the satellite strategy because they are less able to defend a calling space or territory, or because their calls are relatively unattractive to females. In other species, such size differences are negligible, and males may become satellites because they have exhausted the energy reserves required to call (see Energy Requirements of Calling, page 26). Whatever the reason, adopting the satellite strategy can sometimes lead to success, allowing males that otherwise might have missed the opportunity to mate.

A "satellite male" Green Treefrog, inches from a calling male

FEMALE CHOICE

A female frog or toad that is ready to lay eggs moves to the vicinity of a calling male, although he is often too preoccupied to notice. Thus, in many species, the female typically initiates mating by touching or moving very close to the male. But how does the female find the male in the first place? And has she actively chosen among the various males in a chorus?

Numerous scientific studies have demonstrated that a female frog or toad almost always homes in on the male's advertisement calls (see Vocal Communcation, page 21), as opposed to finding him by sight, smell, or other cues. As we might suspect, females also show a strong preference for the calls of their own species. This helps them avoid mating with another species, which may result in no offspring, less viable or sterile offspring, or offspring that are at an ecological disadvantage.

Of additional significance is the fact that females may choose males that are physically or genetically more fit than others. The likely mechanism is that females pick males who are expending the most energy to produce their advertisement calls, meaning those with the longest calls or calls with higher than average repetition rates. Thus, a female may be attracted not only to species-identifying aspects of a male's calls, but also to aspects of his call that are measures of his quality or fitness.

SOUND PRODUCTION

Frogs and toads in North America produce advertisement and aggressive calls by passing a stream of air from the lungs over the vocal cords during expiration. The frog's nostrils and mouth are closed, and there is almost always a vocal sac or sacs that inflate while a call is made.

The vocal sac serves to spread the sound produced by the vocal cords over a larger area, increasing the efficiency of sound transfer to the surrounding air or water. But the vocal sac does not amplify the sound in the sense of adding energy, nor does sound resonate within the

A "knot" of Western Toads (four males and a female in the center)

sac, as many people think. If that were the case, the frequency of the sound would decrease as the sac inflated, which is not usually what occurs.

In some species, such as the Spring Peeper, each call results from a single brief expulsion of air from the lungs and a simultaneous inflation of the vocal sac. The sac collapses between calls at least partially, as air is moved back into the lungs. A call made in this way is accompanied by a sudden contraction of the midsection.

In other species, such as some trilling toads, the sac remains inflated for the duration of the call as the powerful muscles in the midsection smoothly contract, producing a continuous stream of air that passes over the vocal cords. In this case, the pulses within a trill are made by mechanisms in the vocal cords themselves. In other species of trilling toads and in the Gray Treefrog, each pulse of a trill is generated by a single contraction of the muscles of the midsection. This results in a pulsed delivery of air to the vocal cords, which vibrate to produce each pulse in the trill.

ENERGY REQUIREMENTS OF CALLING

The ability to produce advertisement calls is important in the life of a male frog. Unless he adopts the satellite tactic as his breeding strategy, the male must use his advertisement call to attract a mate. It may not be obvious, but producing such calls requires a considerable output of energy. In fact, in some species of frogs, calling causes the frog's metabolic rate to increase ten to twenty times more than its resting rate, which is more energy costly than if the frog hopped nonstop until exhausted. In order to sustain continued calling, the heart and lungs must work at full bore; in species that breed for weeks or months, calling males lose significant proportions of their body weight.

HEARING

Most frogs and toads have prominent circular eardrums, or tympanic membranes, which are positioned just behind the ears. These membranes vibrate in response to sound, and the vibrations are passed to two inner-ear organs, the amphibian and basilar papillae, the former being a diagnostic characteristic of the class Amphibia. In general, within a species there is a fairly good match between the frequencies to which these organs are most sensitive and the frequencies that are predominant in that species' advertisement call. Such tuning, which is referred to as matched filtering, may improve an individual's ability to perceive the calls of its own species in a cacophonous mixed-species chorus.

Unrelated to the hearing organs is a phenomenon called temperature coupling, which was described many years ago in another cold-blooded group, the tree crickets. While the frequency of a male's call is only weakly affected by temperature, the temporal aspects of his call, such as pulse rate within a trill, may be strongly influenced by it, with pulse rates generally speeding up as the temperature increases. Studies of the Gray Treefrog and Cope's Gray Treefrog show that temperature has a parallel effect on the central nervous system of females, with cool females preferring the slower pulse rate trills produced by cool males, and warm females preferring the faster pulse rate trills of warm males.

Gray Treefrog

Evolution and Speciation

In this book, we feature 101 species of frogs and toads. Most fit the standard, classic definition of *species* in that each group's members share a distinct look, sound, and genetic makeup. Furthermore, the species generally do not interbreed; when they do, the resulting hybrid offspring are not viable, not fertile, or, at the very least, at a clear disadvantage in coping with their environment. Although about 90 percent of our frogs and toads are considered to be "good species" by anyone's definition, a number of cases are not as clear-cut. These demonstrate the difficulties that scientists face as they work toward understanding the evolution and speciation of the group. So let's consider a few of these cases and the processes that they reveal.

SOME INSTRUCTIVE EXAMPLES

Green Treefrogs and Barking Treefrogs are two clearly different species that overlap in range in certain parts of the Southeast. They look and sound different from each other—they're easy to tell apart by their body proportions, skin texture, and advertisement calls. But despite these obvious physical differences, striking similarities in their DNA have led scientists to conclude that in the distant past they were probably one species. In areas of overlap, the two species tend to be separated by breeding-habitat differences. Yet in rare cases they do hybridize, and the resulting offspring are viable and fertile—a clear deviation from the classic definition of *species.* The conclusion is that they are different species separated by their habitat differences and maintaining their distinct morphology and behavior, but they are also closely related and so are able to produce fertile hybrid offspring.

Now let's consider the Gray Treefrog and Cope's Gray Treefrog, whose ranges overlap in many areas. They look identical and have very similar calls (although the calls do differ in distinct ways). For more than a century, they were considered to be variations of a single species. But then cellular studies revealed something startling: the Gray Treefrog has twice as many chromosomes as Cope's Gray Treefrog and other native treefrogs. Furthermore, interbreeding (which is rare) produces hybrid offspring that are less viable and sterile and that have an intermediate number of chromosomes. Clearly, these are two different species, even though they look identical and sound very much alike.

Now consider the Ornate Chorus Frog and Strecker's Chorus Frog, whose ranges do not overlap and are separated by hundreds of miles. While they can be told apart by appearance, they are nonetheless very similar in body proportions, color patterns, and advertisement calls. They also have similar genes and are thought to have evolved from a common ancestor. This means that we have to guess about whether or not their differences would prevent interbreeding if their ranges were to expand and overlap. But right now their ranges don't overlap and they cannot interbreed. The take-home message is that we consider these two frogs to be separate species because we can reliably tell them apart and there is no possibility of gene exchange.

The plot thickens when we consider Strecker's Chorus Frog and, more specifically, the geographically isolated populations in the northern part of its range. These populations were first considered to be subspecies of the more wide-ranging form and were given the name

Illinois Chorus Frog. Because of their isolation and subtle differences in appearance, the Illinois Chorus Frog was later considered by some authorities to be a full species. Then another in-depth molecular study appeared supporting their original designation as subspecies. It's a little confusing. Although there are only minor differences in appearance, calls, and DNA, the two subspecies are geographically separated, meaning that they are not currently exchanging genes. Presumably, they are adapting to their different environments and may eventually differ enough from each other for each group to warrant the status of species. Thus, they are subspecies now but may be considered separate species in the distant future.

These examples reveal the reasoning behind decisions that scientists make about what constitutes a species. It is not all black and white, and as you might imagine, biologists quarrel about such things. New studies that reverse previous decisions continue to appear.

MODES OF SPECIATION

In the previous section we explained that Green and Barking Treefrogs evolved from a common ancestor. And so did the Ornate and Strecker's Chorus Frogs. There is strong genetic evidence for this. But how did such speciation come about? What are the primary mechanisms by which most of our North American frogs and toads evolved?

One of the most common and least controversial modes of speciation involves the emergence of a geographical barrier that splits a single species into two or more different groups. This geographical barrier may result from the formation of a mountain range, the expansion of an ocean, the creation of a desert, or the intrusion of a glacier. In the two examples mentioned, Miocene inundations by the sea into the region now occupied by

A male American Toad mounted on top of a female Fowler's Toad — hybrid offspring produced by these closely related species will be fertile.

the Mississippi River (as far north as its junction with the Ohio River) almost certainly played a major role. Eastern and western populations of the ancestral species became separated by this saltwater barrier and gradually adapted to their different environmental conditions.

Thus, one ancestor differentiated into Ornate and Strecker's Chorus Frogs, and another ancestor gave rise to Green and Barking Treefrogs. While the former two species maintained separate ranges after the sea receded, Green Treefrogs expanded their range from west to east so that they now coexist with Barking Treefrogs throughout much of the Southeast.

Other species, such as those in what is termed the western toad complex (Western Toad, Wyoming Toad, Black Toad, Amargosa Toad, and California Toad), were undoubtedly created by climate change and the expansion of deserts, which left behind islands of suitable habitat where there was enough water for the toads to reproduce. In each of these desert oases, the animals experienced different mutations and adapted to different environmental challenges, and in the process they became separate species. To this day they remain separated, but they could come into contact in the distant future.

CONTACT ZONES

If barriers that separate species are removed so that they come into contact, there are three basic possible outcomes: (1) they maintain their genetic integrity and coexist as clearly identifiable species; (2) they interbreed to the extent that their differences become reduced; or (3) they coexist or hybridize wherever their newly expanded ranges meet, but these zones remain very narrow and species differences are largely maintained.

The first outcome is exemplified by the Green and Barking Treefrogs, which rarely hybridize. In areas of overlap, they are able to coexist because they have adapted to different habitats.

The second outcome is demonstrated by the Green Frog. Individuals in the southern portion of the species' range are called Bronze Frogs and have plain brown or bronzy backs, in contrast to northern individuals that have green or greenish-brown backs. These two forms were probably once separated geographically but now have a broad area of overlap in which they interbreed freely to produce fertile offspring with intermediate physical characteristics. Because of extensive gene exchange, the two are considered subspecies, and differences between them may eventually disappear.

A male Squirrel Treefrog mounted on top of a female Barking Treefrog—these two species are genetically related and may produce offspring, but the offspring will probably have low viability.

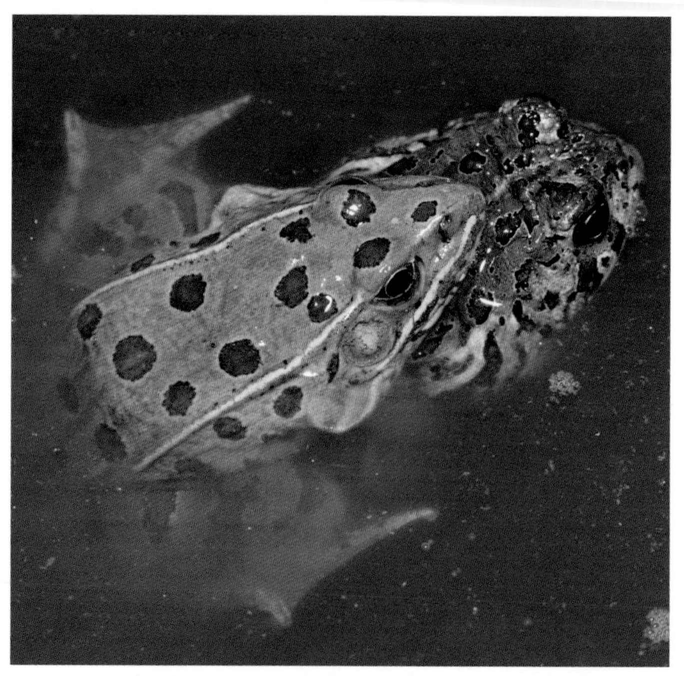

A male Plains Leopard Frog mounted on top of a female American Toad—these two species are genetically incompatible and no offspring will result from this pairing.

Upland Chorus Frog

The third outcome, where zones of contact remain narrow and species differences are maintained, is a complex situation that is illustrated in North America by a number of closely related leopard frogs, and especially by a group of chorus frogs that include the Western, Upland, Midland, New Jersey, and Cajun. Until fairly recently, several of these chorus frogs were considered to be subspecies. But then differences in their calls, subtle differences in appearance and anatomy, and mild to severe genetic incompatibility were documented. Now they are all considered separate species, most with narrow ranges of overlap and limited gene exchange. Nonetheless, the situation is dynamic, and it is possible that one or more of these species is gradually replacing another. The reason is that, unlike the Barking and Green Treefrogs (see above), these species have not evolved differences in ecology that allow them to coexist without competing for food, shelter, and other resources.

OTHER TAXONOMICAL CHALLENGES

The assignment of the genus to a species can sometimes be as problematic as defining a species. For example, recent popular field guides have used three different generic names—*Pseudacris, Hyla,* and *Limnoedus*—for the Little Grass Frog. The current designation is *Pseudacris ocularis,* which embraces the generic name that was used for the species nearly a century ago. Such changes often result from broad and complicated taxonomic studies that take into account a multiplicity of variables, including anatomy and genetics.

Taxonomic classification is always in a state of flux. For example, North American toads that have long been considered to be in the genus *Bufo* are now considered by certain authorities to be sufficiently distinct from their European and African relatives to be placed in an entirely new genus, called *Anaxyrus.* Likewise, certain members of the genus *Rana* may soon be classified as *Lithobates,* setting them apart from other members of their family. While these new names may better reflect evolutionary relationships, their adoption will no doubt have a disruptive effect on communications among scientists in different disciplines, where the old names are well ingrained. Thus, the genera *Bufo* and *Rana* are likely to persist in scientific literature long after field guides have adopted new names. We may take comfort in thinking that classification is static and unchanging, but the reality is that taxonomy is a science, and names must therefore change in response to new ways of interpreting existing information as well as to new discoveries about the animals themselves.

Conservation Issues

In the history of life on Earth there have been five mass-extinction events, in which the majority of then-existing species perished. The most recent of these was sixty-seven million years ago, when the dinosaurs and many other life forms died out, most likely as a result of an asteroid striking Earth. We are now in the midst of a sixth great extinction wave, but unlike the previous five, this one is caused by human activities, especially humans' destruction and alteration of habitats. Around the globe almost all groups of organisms are suffering declines as part of the biodiversity crisis. Yet, even within this sixth wave, something worse and unique is happening to amphibians.

Amphibians around the world, including North American frogs and toads, are facing an extinction crisis that is unprecedented in human history. Never before have we seen an entire class of vertebrates—frogs and toads, salamanders, and a little-known group called caecilians—so threatened by population declines and mass extinctions. According to the first global assessment of the status of amphibian species, more than 40 percent of the world's almost six thousand amphibian species have experienced recent declines, and nearly a third are threatened with extinction. This is far worse than comparable assessments for birds and mammals.

While not all amphibian species are threatened, rapid population declines have been documented around the planet: in the United States, in Central and South America, in Australia and New Zealand, in Europe, and in some places in Africa (we currently don't know enough to assess declines in Asia). In many locations, entire populations have disappeared from one year to the next. If a species is confined to a few areas, such rapid die-offs can easily lead to the species' extinction. A well-known example is the Golden Toad, which was abundant in the Monteverde Cloud Forest Preserve in Costa Rica until 1987. By 1989, only a few animals remained, and, since that time, no individuals have been seen. If a species is lucky enough to have a large range, populations may disappear from parts of their range while persisting in other parts. For example, the California Red-legged Frog has disappeared from approximately 70 percent of its range, yet the remaining populations on California's central coast are abundant and apparently healthy.

CAUSES OF AMPHIBIAN DECLINES

A striking feature of amphibian population declines, and what sets them apart from the general biodiversity crisis, is that they have occurred in some of our best-protected natural areas. For example, Sierra Nevada Yellow-legged Frogs have disappeared from wilderness areas high in the Sierra Nevada. While remote locations such as these have escaped obvious habitat alteration and destruction, they are still being affected by human activities. For the last fifteen years, scientists have investigated five possible causes for the mysterious rapid amphibian declines that have occurred in such remote locations: (1) introduced nonnative species, (2) increased ultraviolet radiation, (3) disease, (4) climate change, and (5) toxic contaminants such as pesticides.

People introduce nonnative species for a variety of reasons, including for recreational fishing, for human food, and for biological control of pests. For example, as early as the 1800s, fish stocking of high-elevation lakes in

Western Toad — severe declines in the southern Rocky Mountains

the Sierra Nevada was done by horseback. Today, trout are raised in hatcheries and dropped by airplane into remote sites in mountain ranges throughout the West. Trout are voracious predators and will eat both tadpoles and adult frogs, so they could easily be the culprit behind some declines. While introduced species such as trout have significant negative impacts on certain amphibian species, scientists are concluding that they are not the predominant cause of rapid declines.

Increased ultraviolet radiation was long thought to play a role in declines, and it has received much scientific attention. The release of ozone-destroying chemicals such as chlorofluorocarbons (CFCs), formerly used in refrigerators, has resulted in increased ultraviolet radiation reaching the earth's surface. Ultraviolet radiation can cause DNA damage and affect animal immune systems. However, the largest number of rapid declines has occurred in tropical montane forests where amphibians are sheltered from sunlight by the forest canopy. Thus, ultraviolet radiation is unlikely to be a primary cause of these declines.

In 1998 scientists discovered a previously unknown fungus that was associated with frog die-offs in both Australia and Panama. The fungus, *Batrachochytrium dendrobatidis,* is in the chytrid family of fungi and is the only member known to be a pathogen to vertebrates. The discovery of the chytrid fungus proved to be a turning point, and now disease, particularly that caused by the chytrid fungus, has emerged as the leading explanation for rapid amphibian population declines. Chytrid fungus has now been found in more than four hundred amphibian species and has been associated with population die-offs in North, Central, and South America, as well as in Africa, Europe, Australia, and New Zealand. While it is clear that chytrid fungus is the immediate cause of many rapid amphibian population declines, there is still much debate about the disease. Is chytrid fungus a newly emerging disease, spreading and attacking defenseless amphibian hosts? Or have recent environmental changes facilitated disease outbreaks by a pathogen that has long been widespread?

Global climate change, especially warming trends, could be contributing to chytrid fungus outbreaks. Un-

California Red-legged Frog — severe declines over much of its range

usually hot spells may stress amphibians, which would weaken their immune systems and make them more susceptible to disease. Another possibility is that temperature shifts caused by climate change may create environments more favorable for the growth of chytrid fungus. In addition, climate change may cause drier conditions, forcing amphibians to congregate in smaller bodies of water, facilitating the transmission of disease. All these factors point to a possible connection between climate change and disease, but to date the necessary research has not been done to determine if any of these links are actually operating.

Pesticides and other contaminants may facilitate disease also. Such chemicals can be carried long distances from where they are applied and often wind up in remote ecosystems. Studies have shown that contaminants can suppress immune systems in many organisms, including amphibians, which may lead to disease outbreaks. In California, researchers have found a strong association between the geographic pattern of population declines of a number of frog species and the pattern of pesticide applications. However, as with climate change, few studies have been conducted to establish a clear link among contaminants, disease, and amphibian declines. More research is clearly needed.

While we are making much progress in understanding amphibian population declines, there are still key questions that remain unanswered. For instance, why do some species and some regions experience major declines while others do not? In the United States, rapid population declines have been concentrated in western mountains and in the desert Southwest, while the eastern half of the country has been spared. It was originally thought that chytrid fungus was present only in the West. However, recent studies indicate that the fungus is widespread in amphibians across the country. So why are chytrid-infected populations in North Carolina and New York able to live with the pathogen while infected frog populations in California and Arizona may experience massive die-offs? Another puzzle is how the chytrid fungus moves. In a high-elevation basin in the Sierra Nevada, the fungus appears to be spreading slowly between lakes, moving less than a mile a year. If it moves that slowly and is a newly emerging disease, how was it able to spread to remote regions all around the globe in such a relatively short time?

DECLINES AND DEFORMITIES

In 1995 a group of middle-school children in Minnesota found a pond with large numbers of deformed frogs—gruesome frogs with multiple extra limbs or with missing limbs. The find captured widespread public attention. Since then, deformed frogs have been found at many locations across the country. Research has revealed that the chief culprit is a parasitic trematode fluke. Agricultural contaminants, in this case runoff containing fertilizers, increase trematode abundance and therefore boost the incidence of deformities. Chemicals such as pesticides may also be playing a role, but to date the research is inconclusive. Despite the similarities between the causes of rapid amphibian population declines and the causes of amphibian deformities, the two issues appear to be unrelated. While deformities are clearly not good for individual frogs, the species that have populations with high incidences of deformities are not experiencing significant declines.

WHAT YOU CAN DO

One of the most important things an individual can do to help ensure that our frogs and toads are around for

Sierra Nevada Yellow-legged Frog—severe declines in the Sierra Nevada

the future is support local land conservation. Although rapid population declines due to disease are affecting many amphibian species, loss of habitat still remains an important long-term threat. And we as individuals can do something about habitat loss. Most amphibians rely on aquatic ecosystems, such as lakes, streams, bogs, and vernal pools, and these ecosystems have been some of the hardest hit by habitat destruction. Preserving your local stream or wetland is one of the best things you can do for frogs and toads.

You can also help amphibians by participating in local amphibian survey and monitoring programs. This involves listening for frogs and toads, identifying species, and reporting your findings. If you want to get involved, there are several options. In the United States, calling surveys by state are coordinated by the North American Amphibian Monitoring Program (NAAMP), which is a collaborative effort of the United States Geological Survey and various regional partners (pwrc.usgs.gov/naamp). The monitoring program is designed to identify and quantify population trends. After an initial training phase, volunteers conduct surveys in which they stop at specified points along roadside routes and assess calls for a set period of time. These are similar to USGS breeding-bird surveys, but they are performed at night. Surveys are currently being conducted in most states east of the Great Plains. Visit the NAAMP website for more information and to volunteer. Two other more entry-level and family-friendly monitoring programs are Frogwatch USA (frogwatch.org) and Frogwatch Canada (frogwatch.ca). These programs gather basic calling information, encourage people of all ages and in all states and provinces to participate, and help volunteers learn to identify the frogs and toads in their surroundings. Some view Frogwatch as a great place to get one's feet wet before becoming involved in the more serious NAAMP survey effort.

Eastern Narrowmouth Toad

How can you help stop rapid population declines? This is more difficult, but there are several things you can do. If you hike, fish, or otherwise recreate in the mountains of the West or Southwest, it is important not to accidentally spread chytrid fungus. Take care not to move frogs or water from one location to another, and if you catch a frog or toad, let it go exactly where you caught it. We encourage you to support organizations that are actively involved in conservation issues concerning frogs and toads, such as Partners in Amphibian and Reptile Conservation (parcplace.org) and the Amphibian Conservation Action Plan (amphibians.org). The Amphibian Conservation Action Plan is the first comprehensive attempt to protect amphibians around the world. The plan outlines research, captive breeding programs, and a variety of other actions to address immediate extinction threats such as chytrid fungus as well as to protect critical amphibian habitats threatened by human activities.

Western Toad

Miscellaneous Explanations

THE COMPACT DISC

The audio compact disc has been carefully crafted to provide the listener with excellent examples of the calls of nearly all the frogs and toads covered in this book. See page 324 for detailed descriptions of the tracks. The recordings for each species are preceded by the common and scientific names. The tracks on the disc correspond to the species profile numbers in the book, as presented in the list on the following pages and in the upper left corner of each species' page spread. Because there are only 99 tracks on the compact disc, and 101 species are covered in the book, the last two tracks (98 and 99) include two species each.

THE RANGE MAPS

The range maps in this book have been drawn with the help of several different references. Our primary source for the United States was the National Amphibian Atlas (website: www.pwrc.usgs.gov/naa) maintained by the USGS Patuxent Wildlife Research Center. These maps, which are periodically updated, are available as an educational tool for the general public and as a source of information for research and habitat management. Please consult the National Amphibian Atlas for detailed views of ranges of the species covered in this book. While these maps are kept fairly up-to-date, they may not reflect the rapid population declines of some western species.

Our primary references for ranges in Canada and northern Mexico were the two Peterson Field Guides to Reptiles and Amphibians (see Sources and Further Reading, page 336). The maps in the western guide summarize population declines. More information was obtained through a variety of regional guides, supplemented by sources on the Internet. While we believe our range maps are reasonably accurate, there may be some errors, especially with regard to rapid population declines.

LENGTHS

The length of a frog or a toad is measured from the tip of the snout to the vent—the legs are not included in the overall measurement. Lengths are given as a range (for example, 1″–2½″) that indicates the expected span of sizes for adults of each species. Nearly all of the lengths in this book were obtained from the two Peterson Field Guides to reptiles and amphibians and supplemented by information provided by various herpetologists.

ORDERING OF SPECIES

Deciding how to order species within each family presented a challenge. While we might have simply alphabetized species by common or scientific name, it was clear that this would result in an artificial order that does little to reflect relationships among frogs and toads, either geographic or by appearance and evolutionary kinship. After careful consideration, we finally agreed upon a strategy. First, we group closely related species that have similar appearances and calls. We also tend to begin with common, widespread, and well-known species before covering those that are uncommon or rare. And finally, we cover eastern species first, then move on to western species. We trust that this approach will satisfy the majority of our readers.

NUMBER OF SPECIES

There are differences of opinion concerning the total number of species of frogs and toads found in North America. In this book, we describe 101 species, which include 4 introduced species: Cuban Treefrog, Greenhouse Frog, Puerto Rican Coqui, and African Clawed Frog. In January 2008, the Society for the Study of Amphibians and Reptiles (SSAR) published their most recent list of the scientific and standard English common names of North American amphibians (see Sources and Further Reading, page 336). The SSAR lists 99 native species, a number that increases to 103 if the 4 introduced species are added.

The discrepancy between our species number (101) and the SSAR species number (103) has several explanations. First, we recognize the newly described Cajun Chorus Frog, which SSAR has not yet considered. Second, SSAR defines the Illinois Chorus Frog as a species, while we consider it a subspecies of Strecker's Chorus Frog. Finally, SSAR supports a split of the Pacific Chorus Frog (Pacific Treefrog) into three separate species. While we believe this split will probably be upheld, we have chosen not to treat them separately, but rather to briefly describe them in the Pacific Chorus Frog account. One reason is that, because these species have been defined primarily on the basis of mitochondrial data, virtually no information is currently available about differences in their appearances and calls. Given that taxonomists inevitably have differing viewpoints concerning these matters, we have done our best to make intelligent decisions, while remaining fully aware that current and future research will make revisions necessary.

NAMING OF SPECIES

Scientific names are always in a state of flux, and some notable changes are in the making. For instance, the 2008 SSAR species list advises several major generic name changes that are receiving mixed acceptance by the scientific community. According to SSAR (and the scientific studies supporting their decisions), the new genus *Anaxyrus* replaces *Bufo* for most North American toads, and the genus *Lithobates* replaces *Rana* for many familiar species, including the Bullfrog, Green Frog, and the various leopard frogs. Other notable changes include placing two of our toads, the Sonoran Desert Toad and the Coastal Plain Toad, in the genus *Ollotis,* and the Cane Toad in the genus *Rhinella.* While we feel it is currently premature to fully adopt these new generic name changes, we alert the reader to the possibility that they will become widely accepted by discussing them in the family introductions and including them in parentheses in the headings of relevant species accounts.

Illinois Chorus Frog subspecies

Species and Track List

The numbers refer to the 101 species profiles and are equivalent to track numbers on the disc. Note, however, that the last four species are grouped on two tracks to stay within the 99-track limitation of compact discs. See page 324 for detailed descriptions of the recordings featured on each track.

Treefrogs and Allies—Hylidae

Hyla—Treefrogs

1. Green Treefrog—*Hyla cinerea*
2. Barking Treefrog—*Hyla gratiosa*
3. Pine Barrens Treefrog—*Hyla andersonii*
4. Squirrel Treefrog—*Hyla squirella*
5. Pine Woods Treefrog—*Hyla femoralis*
6. Gray Treefrog—*Hyla versicolor*
7. Cope's Gray Treefrog—*Hyla chrysoscelis*
8. Bird-voiced Treefrog—*Hyla avivoca*
9. Arizona Treefrog—*Hyla wrightorum*
10. CanyonTreefrog—*Hyla arenicolor*

Pseudacris—Chorus Frogs

11. Spring Peeper—*Pseudacris crucifer*
12. Little Grass Frog—*Pseudacris ocularis*
13. Ornate Chorus Frog—*Pseudacris ornata*
14. Strecker's Chorus Frog—*Pseudacris streckeri*
15. Midland Chorus Frog—*Pseudacris triseriata*
16. Upland Chorus Frog—*Pseudacris feriarum*
17. New Jersey Chorus Frog—*Pseudacris kalmi*
18. Cajun Chorus Frog—*Pseudacris fouquettei*
19. Boreal Chorus Frog—*Pseudacris maculata*
20. Southern Chorus Frog—*Pseudacris nigrita*
21. Brimley's Chorus Frog—*Pseudacris brimleyi*
22. Mountain Chorus Frog—*Pseudacris brachyphona*
23. Spotted Chorus Frog—*Pseudacris clarkii*
24. Pacific Chorus Frog—*Pseudacris regilla*
25. California Chorus Frog—*Pseudacris cadaverina*

Acris—Cricket Frogs

26. Northern Cricket Frog—*Acris crepitans*
27. Southern Cricket Frog—*Acris gryllus*

Osteopilus—West Indian Treefrogs

28. Cuban Treefrog—*Osteopilus septentrionalis*

Smilisca—Mexican Treefrogs

29. Lowland Burrowing Treefrog—*Smilisca fodiens*
30. Mexican Treefrog—*Smilisca baudinii*

True Toads—Bufonidae

Bufo—True Toads

(see page 126 concerning *Anaxyrus* and other new genera)

31. American Toad—*Bufo americanus*
32. Fowler's Toad—*Bufo fowleri*
33. Woodhouse's Toad—*Bufo woodhousii*
34. Southern Toad—*Bufo terrestris*
35. Oak Toad—*Bufo quercicus*
36. Coastal Plain Toad—*Bufo nebulifer*
37. Texas Toad—*Bufo speciosus*
38. Houston Toad—*Bufo houstonensis*
39. Cane Toad—*Bufo marinus*
40. Great Plains Toad—*Bufo cognatus*
41. Canadian Toad—*Bufo hemiophrys*
42. Wyoming Toad—*Bufo baxteri*
43. Red-spotted Toad—*Bufo punctatus*
44. Green Toad—*Bufo debilis*
45. Sonoran Green Toad—*Bufo retiformis*
46. Sonoran Desert Toad—*Bufo alvarius*
47. Arizona Toad—*Bufo microscaphus*
48. Western Toad—*Bufo boreas*
49. Amargosa Toad—*Bufo nelsoni*
50. Black Toad—*Bufo exsul*
51. Arroyo Toad—*Bufo californicus*
52. Yosemite Toad—*Bufo canorus*

True Frogs — Ranidae

Rana—True Frogs

(see page 184 concerning the new genus *Lithobates*)

53. Bullfrog — *Rana catesbeiana*
54. Green Frog — *Rana clamitans*
55. Pig Frog — *Rana grylio*
56. Mink Frog — *Rana septentrionalis*
57. Wood Frog — *Rana sylvatica*
58. Carpenter Frog — *Rana virgatipes*
59. Florida Bog Frog — *Rana okaloosae*
60. River Frog — *Rana heckscheri*
61. Crawfish Frog — *Rana areolata*
62. Gopher Frog — *Rana capito*
63. Dusky Gopher Frog — *Rana sevosa*
64. Pickerel Frog — *Rana palustris*
65. Northern Leopard Frog — *Rana pipiens*
66. Southern Leopard Frog — *Rana sphenocephala*
67. Plains Leopard Frog — *Rana blairi*
68. Rio Grande Leopard Frog — *Rana berlandieri*
69. Relict Leopard Frog — *Rana onca*
70. Chiricahua Leopard Frog — *Rana chiricahuensis*
71. Lowland Leopard Frog — *Rana yavapaiensis*
72. Tarahumara Frog — *Rana tarahumarae*
73. Cascades Frog — *Rana cascadae*
74. Northern Red-legged Frog — *Rana aurora*
75. California Red-legged Frog — *Rana draytonii*
76. Oregon Spotted Frog — *Rana pretiosa*
77. Columbia Spotted Frog — *Rana luteiventris*
78. Foothill Yellow-legged Frog — *Rana boylii*
79. Sierra Madre Yellow-legged Frog — *Rana muscosa*
80. Sierra Nevada Yellow-legged Frog — *Rana sierrae*

North American Spadefoots — Scaphiopodidae

Scaphiopus — Southern Spadefoots

81. Eastern Spadefoot — *Scaphiopus holbrookii*
82. Hurter's Spadefoot — *Scaphiopus hurterii*
83. Couch's Spadefoot — *Scaphiopus couchii*

Spea — Western Spadefoots

84. Plains Spadefoot — *Spea bombifrons*
85. Great Basin Spadefoot — *Spea intermontana*
86. New Mexico Spadefoot — *Spea multiplicata*
87. Western Spadefoot — *Spea hammondii*

Microhylid Frogs and Toads — Microhylidae

Gastrophryne — North American Narrowmouth Toads

88. Eastern Narrowmouth Toad — *Gastrophryne carolinensis*
89. Great Plains Narrowmouth Toad — *Gastrophryne olivacea*

Hypopachus — Sheep Frogs

90. Sheep Frog — *Hypopachus variolosus*

Neotropical Frogs — Leptodactylidae

Eleutherodactylus — Rain Frogs

91. Greenhouse Frog — *Eleutherodactylus planirostris*
92. Puerto Rican Coqui — *Eleutherodactylus coqui*
93. Cliff Chirping Frog — *Eleutherodactylus marnockii*
94. Spotted Chirping Frog — *Eleutherodactylus guttilatus*
95. Rio Grande Chirping Frog — *Eleutherodactylus cystignathoides*

Craugastor — Robber Frogs

96. Barking Frog — *Craugastor augusti*

Leptodactylus — Neotropical Grass Frogs

97. White-lipped Frog — *Leptodactylus fragilis*

Tailed Frogs — Ascaphidae

Ascaphus — Tailed Frogs

98a. Rocky Mountain Tailed Frog — *Ascaphus montanus*

98b. Pacific Tailed Frog — *Ascaphus truei*

Burrowing Toads — Rhinophrynidae

Rhinophrynus — Burrowing Toads

99a. Mexican Burrowing Toad — *Rhinophrynus dorsalis*

Tongueless Frogs — Pipidae

Xenopus — Clawed Frogs

99b. African Clawed Frog — *Xenopus laevis*

Treefrogs and Allies — Family Hylidae

The treefrog family, Hylidae, is large and diverse. Worldwide, there are about 800 species and nearly 45 genera. In the United States and Canada alone, there are 30 species represented by 5 different genera. The term *treefrog* is generally used to describe slim-waisted, long-legged frogs with sticky toe pads that enable them to live in trees and shrubs. However, not all members of the family are referred to as treefrogs, and many lack well-developed toe pads and do not climb trees. For this reason, we prefer using the word *hylid* to refer in general to any member of the family.

In North America, the hylids that we commonly call treefrogs are members of the genus *Hyla*. There are ten species, and nearly all live in the East, and more particularly the Southeast. They exhibit the classic treefrog body plan and use their large toe pads to climb in bushes, shrubs, and trees. Some can easily change colors. Their calls are varied, ranging from hoarse, nasal calls (Green Treefrog) to musical trills (Gray Treefrog) and birdlike whistles (Bird-voiced Treefrog).

Southern Cricket Frog

Cricket frogs in the genus *Acris* make up another group in the family Hylidae. They are named for their cricketlike calls. There are two species in the United States: the Northern Cricket Frog and the Southern Cricket Frog. Confined mostly to the eastern half of the country, they are small and slender frogs that live near permanent water. Their toe pads are negligible and they show no arboreal tendency. Cricket frogs have a characteristic way of appearing to skip across the surface of the water, seeming to disappear when they come to rest because they are so well camouflaged.

Chorus frogs of the genus *Pseudacris* are a diverse group of hylids. There are fifteen species in North America. Most have reduced toe pads and exhibit terrestrial habits outside the breeding season. Spring Peepers are by far the most familiar chorus frogs in the eastern United States. The Pacific Chorus Frog (Pacific Treefrog), common in the Pacific coastal regions, is universally heard but often not recognized — its familiar *rib-bit* call can be heard in countless Hollywood films and has become the archetype of how a frog should sound. Most other species are secretive and unlikely to be encountered unless they are tracked down by their calls during the breeding season. The group also includes the Little Grass Frog, one of the smallest vertebrates in North America.

Two other genera of hylids are found in North America. The genus *Smilisca* contains two species that are primarily tropical in distribution and that barely range into the United States: the Lowland Burrowing Treefrog of southern Arizona and the Mexican Treefrog of the lower Rio Grande Valley in Texas. Our sole representative of the genus *Osteopilus* is the Cuban Treefrog, an introduced species found in Florida.

Cope's Gray Treefrog

Green Treefrog

Hyla cinerea (1 1/4″–2 1/2″)

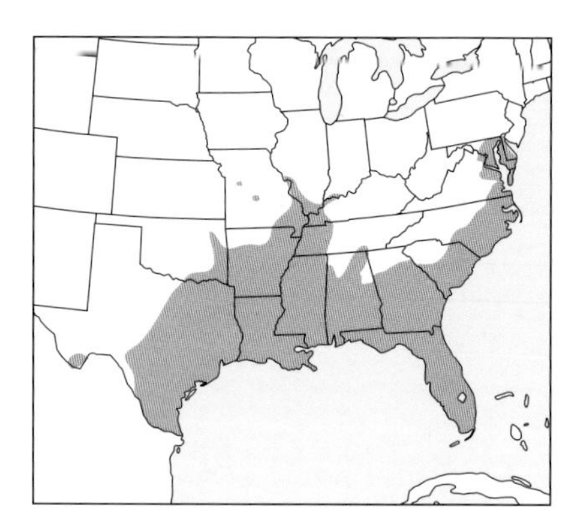

Bright and attractive, the Green Treefrog is a common resident of swampy areas throughout the southeastern and Gulf Coast states. It is often called rain frog, because huge choruses erupt after warm rains, or cowbell frog, because distant calls can have a bell-like quality. The nasal calls of the species, emanating from swamps and bayous from spring through summer, are perhaps the most familiar of all the frog sounds of the region.

Appearance: Green Treefrogs are easily identified by their smooth, lime green skin and broad white stripes (usually with black borders) that extend along each side (stripes are absent in some populations in the vicinity of Washington, D.C., and occasionally in individuals elsewhere). Many individuals have yellowish spots on their back. Cold individuals may turn dark olive in color.

Range and Habitat: Found throughout much of the South, ranging north along the coast to Delaware and following the Mississippi River as far north as southern Illinois. Breeds in permanent lakes, ponds, swamps, and ditches, and sometimes in temporary water, especially after heavy rains. During the day, and at any time outside the breeding season, adults frequent bushes and trees not far from breeding sites.

Behavior: Green Treefrogs are prolonged breeders that can be heard calling on warm nights (usually above 65°F) from late March to early August in the southern part of the range. Calling males do not defend a territory, but individuals may call from the same place for several nights in a row. A male often alternates his calls with those of his closest neighbor and switches to distinctive rattling aggressive calls when a rival gets too close. Females usually initiate mating, but a male will stop calling and pursue a female when he detects her approach. This behavior probably helps prevent noncalling satellite males (see page 24) from intercepting females.

Individuals may lack spots on the back, and some populations do not have white stripes on the sides of the face.

Voice: The advertisement call is a short, nasal *quonk* or *quank* given once or twice per second. Neighboring males often alternate calls. The aggressive call is a hoarse, guttural *quarr-quarr-quarr* that is harsher than the advertisement call and repeated more quickly. Rain calls sound much like advertisement calls and are commonly given from bushes and trees.

Barking Treefrog

Hyla gratiosa (2″–2 3/4″)

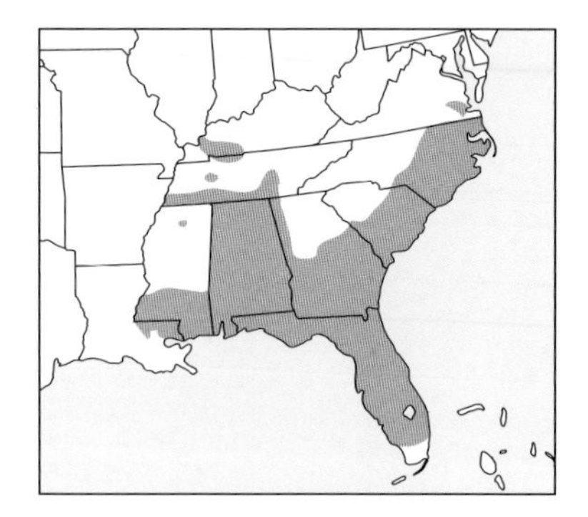

One of our largest native treefrogs, the rotund and attractive Barking Treefrog is a resident of the southern coastal plains. It is named for its barklike rain calls and is easily recognized by its large size, stout build, spotted back, and somewhat warty skin. During the day, Barking Treefrogs climb trees, where they shelter in holes and crevices. They may burrow underground during dry periods.

Appearance: The ground color varies from bright green to dull brown, and individuals are able to change color. Many dark circular or squarish spots are usually evident, although some individuals may be uniformly colored. Usually, ragged or ill-defined whitish stripes lie along each side of the body.

Range and Habitat: Although primarily a coastal plain species, Barking Treefrogs are found as far north as Delaware, and there is an isolated population in western Kentucky. They breed in fish-free bodies of water, including semipermanent ponds, swamps, and ditches in otherwise well-drained areas.

Behavior: Barking Treefrogs are prolonged breeders, breeding from late March until mid-August, depending on rainfall. Males usually call while floating and quickly dive when disturbed. Neighboring males alternate calls. A female nearly always initiates pairing by touching a calling male. Barking Treefrogs hybridize occasionally with Green Treefrogs. Hybrids are usually found in disturbed environments, where sparse vegetation makes it likely that Green Treefrog males and their satellites will be found on the ground. Thus, female Barking Treefrogs hopping toward a breeding chorus are sometimes intercepted by Green Treefrog males. Hybrids may mate successfully with one another or with Green or Barking Treefrogs, but survivorship of hybrids appears to be poor.

Voice: The advertisement call is a resonant *tonk,* repeated about once per second. Rain calls produced from trees or bushes sound similar to the barks of a dog, which accounts for the common name of this species—they are usually repeated more slowly and irregularly than the calls produced in breeding ponds. The pitch of hybrid calls is intermediate between that of Barking and Green Treefrogs.

Pine Barrens Treefrog

Hyla andersonii (1 1/8″–2″)

Perhaps the most beautiful of our native frogs, the Pine Barrens Treefrog is rarely seen and has a restricted and piecemeal distribution, found only in the pine barrens of New Jersey and comparable patches of suitable habitat scattered across the Southeast. By far the best way to find and observe one is by homing in on calling males during the breeding season.

Appearance: The ground color is deep green, bordered on the sides of the head and body with a narrow white line and a broader band of lavender-brown, which extends forward to the nostrils. The hidden parts of the legs and groin are washed in yellow-orange. Superficially resembling the Green Treefrog, Pine Barrens Treefrogs are smaller and stouter in build.

Range and Habitat: These treefrogs breed around bogs and swamps in the pine barrens of New Jersey, from which they get their name. They also inhabit the boggy pocosins—that is, swamps in upland coastal regions—of North Carolina and South Carolina, and sandy bogs of the Panhandle of Florida and adjacent Alabama. They prefer moist, shrubby patches that occur in otherwise very dry habitats such as pinewoods and sandhills.

Behavior: Pine Barrens Treefrogs are spring and summer breeders, and their calling activity is stimulated by rainfall. They call from bog or pond edges, ditches, wet seeps, and semipermanent small streams, and from bushes and small trees near such bodies of water. In some localities in Florida, hybridization between Pine Barrens Treefrogs and both Green Treefrogs and Pine Woods Treefrogs has been reported. (Playback experiments have shown that receptive Pine Barrens Treefrog females respond well to the calls of the similar-sounding Green Treefrog.)

Voice: The advertisement call is a repeated *quonk-quonk-quonk-quonk.* It sounds similar to the call of the Green Treefrog but is slightly higher in pitch and more melodic (not as guttural) and is typically given at a more rapid rate (up to three times per second, compared with about twice per second in Green Treefrogs). Calling is contagious, and small groups will suddenly erupt when one male begins to call after a long period of silence.

Squirrel Treefrog

Hyla squirella (7/8″– 15/8″)

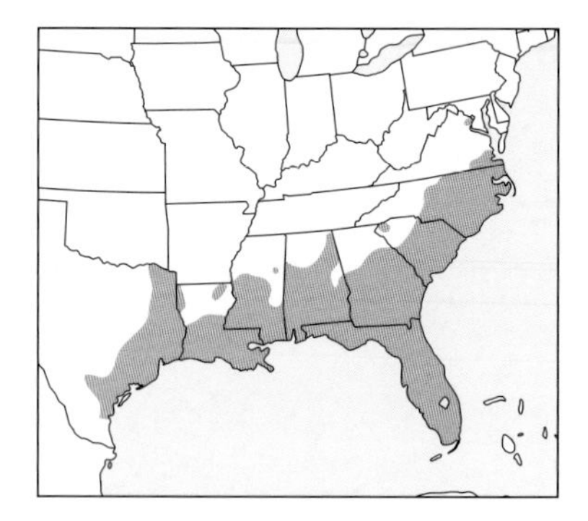

The highly variable Squirrel Treefrog is a common rain frog of the South, and it often produces raspy, squirrellike calls from bushes and trees even before a thunderstorm begins. Squirrel Treefrogs frequent suburban gardens and brushy tangles, where they take shelter in recesses of buildings, in clothesline poles, under boards and pots, and in other moist, concealed areas.

Appearance: The dorsal color ranges from brown to rich green, and spots may be present or absent. Individuals often change in color and pattern over time. There is usually a vague light stripe along each side of the body, but this is never as bold and distinct as the stripes of its larger relative the Green Treefrog. Brown individuals are easily confused with the similar-sized Pine Woods Treefrog, but this latter species has bright white or yellow spots on the backs of its thighs that are lacking in the Squirrel Treefrog.

Range and Habitat: Squirrel Treefrogs occur from Virginia to Texas in coastal plain regions and in peninsular Florida. They occupy a wide variety of habitats, including deciduous woods, pine flatwoods, and sandhills.

Behavior: Squirrel Treefrogs are mainly explosive summer breeders that form large choruses lasting only a night or two. The males congregate and call excitedly from temporary ponds, pools, and roadside ditches. (In many cases, however, these bodies of water dry out before the tadpoles have a chance to transform.) Even when there has been little rain, isolated males may be heard calling in the shallow parts of semipermanent and permanent ponds. Satellite behavior (page 24) is common, and a large male may have several satellites positioned near him, all attempting to intercept females that are attracted to his loud calls.

Voice: The advertisement call is a nasal, buzzing quack that is briskly repeated around two times per second: *rrraak-rrraak-rrraak-rrraak-rrraak*. The Squirrel Treefrog gets its name from another call, a squirrellike rain call that it gives from shrubs and trees away from breeding areas. The rain call is weaker and raspier than the advertisement call and is reminiscent of the scolding notes of gray squirrels.

A male brushing mosquitoes off its head

Pine Woods Treefrog

Hyla femoralis (1″– 1¾″)

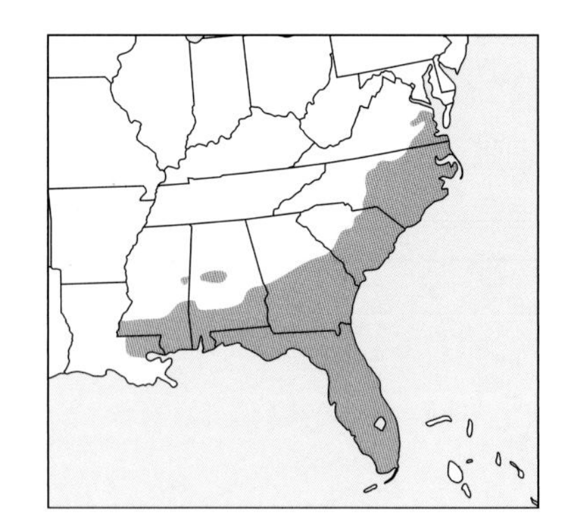

A small and rather nondescript treefrog of the southeastern coastal plain, the Pine Woods Treefrog lives in pine forest and has a unique call that resembles the sound of someone madly tapping out dots and dashes on a telegraph key—some even refer to this anuran as the Morse-code frog. When not breeding, it usually remains high in the trees, but it may also be found foraging in low shrubs and sometimes on the ground.

Appearance: The ground color varies from gray or greenish gray to brown but more commonly is reddish brown with large dark blotches. It is easily confused with brownish Squirrel Treefrogs, which are about the same size and often breed at the same time and in the same place. Fortunately, the calls of the two species are different. Pine Woods Treefrogs can also be identified by the row of small white, yellow, or orange spots found on the hidden parts of the thigh (spots are absent in the Squirrel Treefrog).

Range and Habitat: Inhabits the coastal plain region of the Southeast and peninsular Florida. Breeding sites are typically located in pine flatwoods—hence the common name—but this species is also found in other habitats that have sandy or well-drained soils, including pocosins and cypress swamps.

Behavior: After a heavy summer rain, Pine Woods Treefrogs form large choruses in temporary roadside ditches and pools. These breeding sites are often shared with Oak Toads, Eastern Narrowmouth Toads, Little Grass Frogs, Squirrel Treefrogs, and Cope's Gray Treefrogs. Within several hours of the rain, a pool can be teeming with hundreds of males, all calling within inches of one another. But the raucous chorus is short-lived, and the same pool may be nearly silent the following night.

Voice: The advertisement call is unique among our treefrogs: a nearly continuous string of raspy notes that sounds like frantic tapping on a telegraph key: *dik-dik-dikadikadikadikadikadika . . .* The tempo of calling is often highly irregular. Pine Woods Treefrogs occasionally give a slow-paced series of calls from trees and shrubs away from breeding pools, frequently during the day.

Gray Treefrog

Hyla versicolor (1 1/4″–2 3/8″)

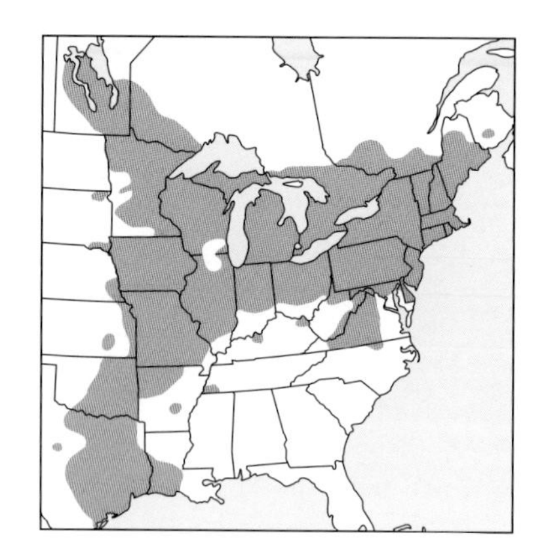

Plump and warty, the Gray Treefrog is commonly discovered around houses surrounded by forest, where it may suddenly call from a concealed location or be found on a potted plant. The scientific name *versicolor* refers to the ability of individuals to change colors, probably to match their surroundings, and this species certainly lives up to the name. Gray Treefrogs are the only polyploid frogs in the United States, meaning that they have twice the number of chromosomes as other treefrogs.

Appearance: The dorsal surface can be light gray, dark gray, brown, or vibrant green—and everything in between—usually accentuated by darker markings of various shapes and sizes. (The accompanying photos show a number of these variations.) There is a distinct light spot beneath the eye, and the hidden parts of the legs and groin are washed with yellow-orange. Gray Treefrogs are indistinguishable in appearance from the closely related Cope's Gray Treefrogs, and the two species can be most reliably told apart by their advertisement calls.

Range and Habitat: Ranges from Texas north to the upper Midwest and then east through most of the Northeast. Distribution is patchy in some areas, especially in Missouri, where populations have not been found in about a third of the counties. (Such extreme patchiness is not included in the range map.) Although generally nonoverlapping with the range of Cope's Gray Treefrog, there are many areas where the two occur together. Daytime retreats are usually in wet wooded areas, where individuals take refuge in tree cavities.

Behavior: Gray Treefrogs are nocturnal and forage in trees and shrubs. They descend to the ground to breed from March to July in semipermanent ponds that lack predatory fish. Calling is energy costly for males, and females prefer to mate with the males that produce the longest, most taxing calls—one study demonstrated that the offspring of males that have long calls survive better and grow faster than those of males with short calls.

Voice: The advertisement call is a melodious trill lasting about half a second and repeated every few seconds. The pulse rate of the trill is about half that of a Cope's Gray Treefrog at the same temperature. Squeaky chirps or *weeps* are given during aggressive encounters between males. Release calls, which are produced by both genders, sound very similar to the aggressive calls.

Cope's Gray Treefrog

Hyla chrysoscelis (1 1/4″–2 3/8″)

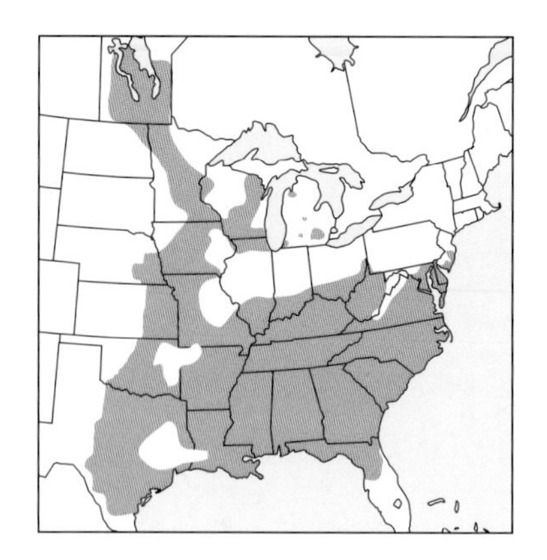

Cope's Gray Treefrog is named in honor or Edward Drinker Cope (1840–1897), an American paleontologist and noted herpetologist. It looks identical to the Gray Treefrog, and the two were originally thought to be a single species. But Cope's Gray Treefrog has the same number of chromosomes as other North American treefrogs, whereas the Gray Treefrog has twice as many. While they cannot be distinguished by their appearance, the two species have noticeably different calls.

Appearance: As in the Gray Treefrog, the ground color ranges from gray to green, and individuals are able to change colors in response to their surroundings. The geographic distribution can make identification easy in some places: Cope's Gray Treefrogs occur through much of the Southeast and lower Midwest, where Gray Treefrogs are not found, and Cope's Gray Treefrogs are notably absent from much of the Northeast, where Gray Treefrogs are common.

Range and Habitat: Cope's Gray Treefrogs are widespread throughout much of the East and are common in the Southeast, Mid-Atlantic, and parts of the Midwest, where they are patchily distributed. They are primarily forest dwellers but may be associated with grasslands and prairies at the western edge of the range.

Behavior: Breeding occurs from March to August in semipermanent, fishless ponds. Males tend to call more often from the ground, rather than in bushes and trees like male Gray Treefrogs. In both species, the trill rate of the advertisement call is strongly influenced by temperature. Warm females prefer the fast rates produced by warm males of their own species, and cool females prefer the slow rates produced by cool males of their own species. Such preferences help keep Gray Treefrogs and Cope's Gray Treefrogs from interbreeding, although hybrids occasionally occur.

Voice: The advertisement call is a rattling trill that is noticeably harsher than the melodic trill of the Gray Treefrog because of its faster pulse rate and pulse characteristics. Calls of the two species are easy to distinguish if males are heard calling at the same time and temperature. Pulse rates of the trills in the western part of the range are higher than those in the East, giving the calls a unique buzzy quality. Males in all areas give chirping or weeping aggressive calls that sound very similar to those of Gray Treefrogs.

Bird-voiced Treefrog

Hyla avivoca (1 1/8"–2 1/16")

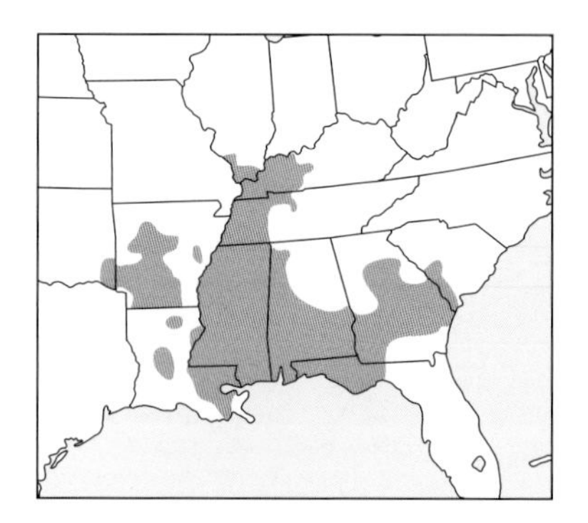

The hauntingly beautiful whistling calls of Bird-voiced Treefrogs echoing from cypress trees in a southern swamp are one of nature's highlights. While one might expect these singers to be as colorful as their songs, the opposite is actually true—they are drab and well camouflaged, and their arboreal habits make them difficult to find.

Appearance: Bird-voiced Treefrogs usually range in color from gray to brown, but are sometimes green. They are easily confused with Gray or Cope's Gray Treefrogs but are generally smaller in size, and the wash on the hidden parts of the thighs and groin is pale green rather than bright orange. Furthermore, the light spot under the eye is sometimes distinctly yellow rather than the whitish color typical of the other two species.

Range and Habitat: A denizen of the great river swamps of the South, ranging from Louisiana to South Carolina and north along the Mississippi River to southern Illinois. Absent from the Florida peninsula. Bird-voiced Treefrogs also breed in large floodplain ponds, man-made ponds, and lakes that are near rivers or streams.

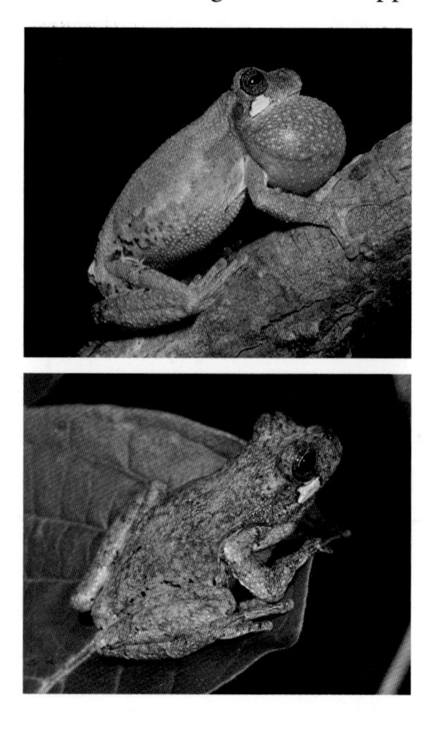

Behavior: Bird-voiced Treefrogs begin calling as early as April in the southern part of their range and continue until early August in most localities. Males call from high in the canopy at the start of the season and also at the beginning of each night's chorusing, then descend to call from perches closer to the water. Breeding males tend to space themselves at greater distances from one another than do other treefrogs, perhaps as a result of their highly aggressive behavior. A male produces distinctive aggressive calls when another male calls nearby, and wrestling matches may ensue, with the victor claiming the calling site, at least for the night.

Voice: The advertisement call is a rapid series of ten to twenty musical whistles lasting several seconds and sometimes varying slightly in tempo. During aggressive encounters, males give a series of harsh, grating trills: *prrreeek, prrrreeek, prrreeek.* Hybrids between Bird-voiced and Cope's Gray Treefrogs often occur, and their calls sound much like the musical trills of Gray Treefrogs.

Arizona Treefrog

Hyla wrightorum (3/4"–2")

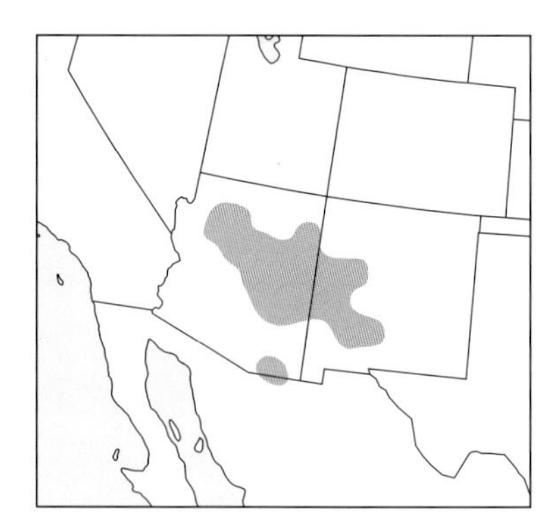

The scientific name of this beautiful frog of the mountains of Arizona and New Mexico honors a famous herpetology couple, Albert and Anna Wright, authors of the classic *Handbook of Frogs and Toads of the United States and Canada,* first published in 1949. Also referred to as Mountain Treefrog, the Arizona Treefrog is common to abundant in many areas, though it is rarely seen outside the breeding season.

Appearance: Arizona Treefrogs range in color from vibrant green or yellow-green to brown. Dark spots or bars may occur on the back, and a dark eye stripe extends past the shoulder onto the side of the body and sometimes to the groin area. Identification is really never a problem because the only other treefrog in the region, the Canyon Treefrog, looks very different.

Range and Habitat: Commonly found in open or semi-open areas, especially wet meadows above 5,000 feet in the ponderosa pine forests of north-central Arizona and west-central New Mexico. Isolated populations (which may soon be categorized as a threatened species) occur in grasslands, pine-oak forests, and mixed conifer forests in the Huachuca Mountains and Canelo Hills of southeastern Arizona.

Behavior: Arizona Treefrogs are explosive breeders, forming noisy choruses after heavy thunderstorms during the summer monsoon season. They breed primarily in ephemeral waters such as roadside ditches, pools that form in meadows, stock ponds, and even in flooded tire tracks. Outside of the breeding season, some individuals probably climb high into trees, but others take refuge on the ground in wet meadows, under logs, and in rock outcrops.

Voice: The advertisement call of the Arizona Treefrog is a brief, guttural quacklike *rawk.* The calls are repeated rapidly, sometimes more than twice a second, and are normally given only at night: *rawk-rawk-rawk-rawk-rawk.*

Canyon Treefrog

Hyla arenicolor (1 1/4"–2 1/4")

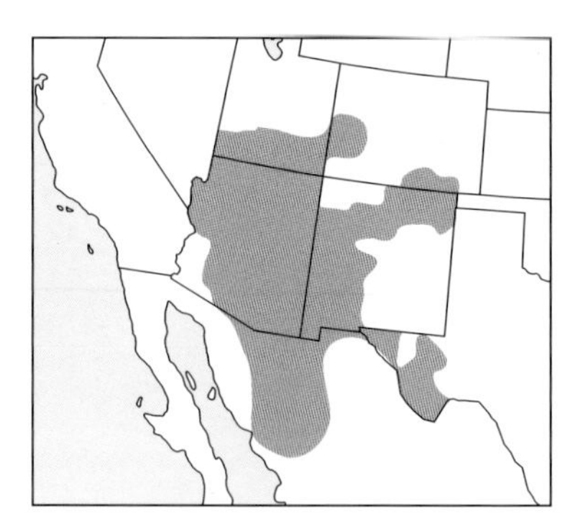

A drab, gray-colored treefrog of the arid Southwest, the Canyon Treefrog is well camouflaged in its rocky environment. It is found along mountain streams, where by day it perches on boulders or hides in rocky nooks and crevices. Canyon Treefrogs can be very difficult to locate, but sometimes individuals stand out against their background, and occasionally one may be found basking in full sun on a rock.

Appearance: The skin is slightly warty, and the ground color is some shade of gray or olive green. There are usually spots on the back, and the toes have large, expanded pads. A wash of yellow or orange occurs under the front and hind legs. The vocal sac is bilobed. This species is similar in appearance to the California Treefrog, but the ranges of the two species do not overlap and their calls are dramatically different.

Range and Habitat: Ranges from Mexico northward through Arizona and New Mexico to Utah, with isolated populations in southern Colorado and western Texas. Canyon Treefrogs are found in and around fast-flowing rocky streams in mountains and foothills, and along sandy streams that meander into the desert from nearby highlands. These treefrogs may also be found along the edges of ponds formed by the damming of a stream.

Behavior: Canyon Treefrogs may breed in the spring, during the summer monsoon rains, or at both times, depending on the locality and rainfall patterns. Spring breeding occurs as early as April in the foothills and in the Grand Canyon, and as late as July at higher elevations in the mountains. Calling is primarily at dusk, but may also occur by day. When males interact, they produce distinctive aggressive calls and sometimes engage in wrestling matches. However, males are not strictly territorial, and they frequently move to new positions in attempts to intercept approaching females.

Voice: The advertisement call is a harsh, rattling, mechanical trill with a pronounced hollow quality: *brrrrrrrr*. It averages about a second long and is repeated every few seconds. Aggressive calls are shorter and sweeter-sounding, with accented endings.

Spring Peeper

Pseudacris crucifer (3/4"– 1 1/2")

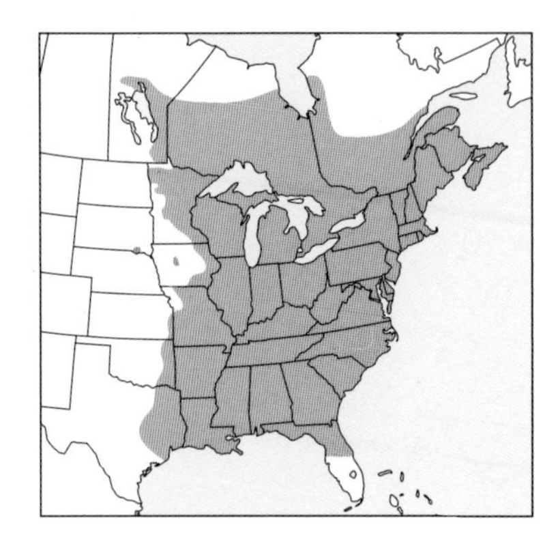

The lovely, nearly pure-tone calls of the diminutive Spring Peeper are among the first frog sounds heard each spring in the eastern United States and Canada. The scientific name *crucifer* refers to the crucifixlike × found on the frog's back. Peeping rain calls given in late summer and autumn have earned this species an alternative and seasonally appropriate name: the Autumn Piper.

Appearance: The dorsal ground color may be straw, brown, gray, olive, or even a rich rusty orange. The dark × marking is usually accompanied by other dark markings, including bands on the hind legs and a line that extends from one eye to the other. Although classified as chorus frogs on the basis of biochemical criteria, Spring Peepers have relatively well-developed toe pads and were formerly considered members of the genus *Hyla*.

Range and Habitat: Common throughout much of the East, although conspicuously absent from most of peninsular Florida, Spring Peepers are primarily woodland frogs. They seldom climb very high, and outside the breeding season they are usually found on the ground and in small bushes near the breeding pond.

Behavior: In early spring, Spring Peepers form deafening choruses in temporary to semipermanent water. Individuals call both day and night as long as the temperature remains above freezing. Calling males often interact to form duets or trios, where they alternate calls in a nonrandom fashion. Males also produce distinctive aggressive calls (see below) when a chorus is just beginning or when rivals approach too closely.

Voice: The advertisement is a nearly pure-tone whistle or *peep* that rises slightly in pitch from beginning to end. It is a loud and piercing call, given about once per second or faster. Distant choruses sound like the jingling of sleigh bells. The aggressive call, given especially during encounters between males, is a stuttering trill, reminiscent of the calls of chorus frogs: *purrrreeeek,* usually rising in pitch at the end. Squeaky peeps (rain calls) are given periodically by individuals from shrubs and trees in late summer and autumn.

12

Little Grass Frog

Pseudacris ocularis (7/16″–11/16″)

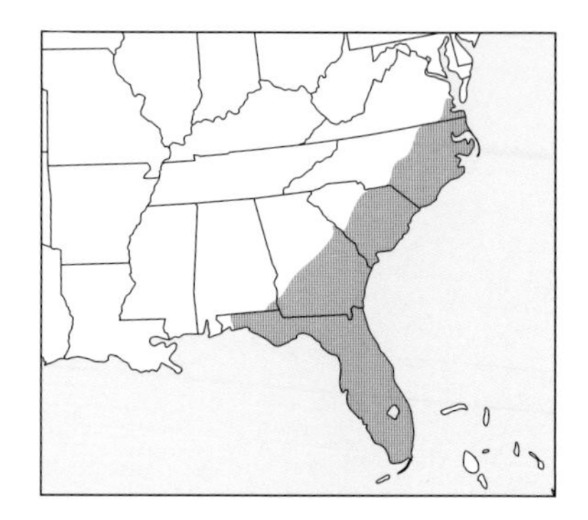

The Little Grass Frog is the smallest frog in North America. Tiny and delicate in appearance, it has well-developed toe pads and was long considered a treefrog of the genus *Hyla*. Little Grass Frogs live up to their name—they frequent open grassy areas and seldom climb higher than the tops of tufts of grass and small shrubs.

Appearance: The dorsal ground color is highly variable, ranging from pale brown or tan to reddish. A reliable identifying feature is the prominent dark stripe (one on each side) that extends from in front of the eye to the side of the body. Less distinctive stripes may occur on the back. The underside is white.

Range and Habitat: Common in the southeastern coastal plain region from Virginia to Florida, including peninsular Florida. Prefers grassy habitats and areas dominated by sedges or sphagnum moss. Often found in cypress swamps, bogs, and along the edges of small ponds. Breeds in wet grassy depressions as well as roadside ditches and temporary pools, where it may be heard calling alongside a number of other species.

Behavior: Little Grass Frogs begin breeding as early as January in the southern parts of the range and may continue through most of the summer. Breeding episodes are triggered by thunderstorms. Although abundant over much of its range, its small size and secretive habits make it inconspicuous.

Voice: The advertisement call, which is the highest pitched of any North American frog (about 5–6 kHz), is an insectlike chirp given about once every second. Each chirp is made up of two distinct parts, a brief introductory note followed by a trill: *pt-zeee.* A chorus sounds like a large group of crickets chirping.

13

Ornate Chorus Frog

Pseudacris ornata (1″– 1 7/16″)

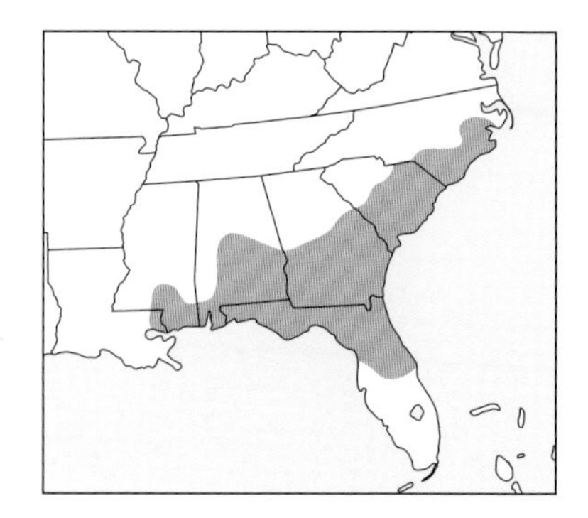

The striking colors and patterning of this chunky chorus frog well justify its common name, although it is a species that is extremely difficult to find and observe outside the breeding season. Frequenting upland areas well away from water, Ornate Chorus Frogs burrow among roots and are seldom seen except when excavated by chance or when they emerge in summer after heavy rains.

Appearance: The ground color can be brown, reddish brown, silvery, or green, and individuals can rapidly change color. There is a prominent black eye stripe that enlarges behind the eye to cover the eardrum and extends to the arm and then along the side, but often discontinuously. There may also be some elongate brown spots or bars on the back. The groin is washed in pale yellow, and there are yellow spots on the hidden parts of the legs.

Range and Habitat: A southeastern species of coastal plains and adjacent piedmont, ranging from North Carolina to extreme eastern Louisiana. Absent from lower Florida. Frequents pine flatwoods, open mixed woods, and other relatively open areas with sandy or otherwise well-drained soils. Breeds in temporary ponds, cypress swamps, roadside ditches, and flooded fields and meadows.

Behavior: Ornate Chorus Frogs are late-fall and winter breeders and do most of their calling from December through March. Males tend to call from more exposed positions than other chorus frogs in the same breeding area, but they are quick to stop calling when disturbed and to dive for cover when approached.

Voice: The advertisement call is a brief, metallic-sounding *peep* or *pip*. It might possibly be confused with the call of a Spring Peeper but is more abrupt and ringing and is given at a much faster rate, two or three notes per second: *pip-pip-pip-pip-pip*. A group of males peeping excitedly is a pleasure to hear.

Strecker's Chorus Frog

Pseudacris streckeri (1″– 1 7/8″)

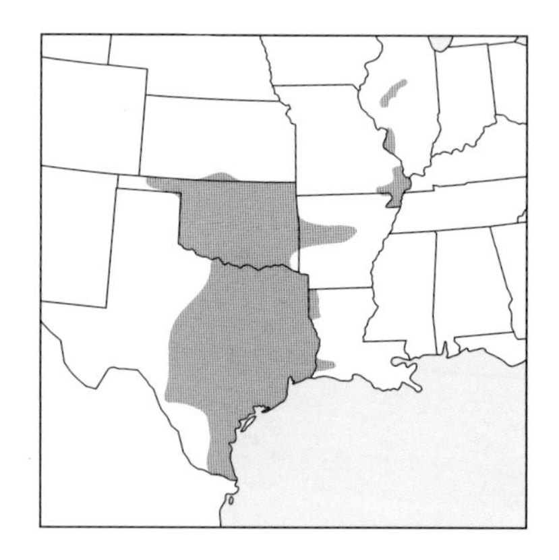

The largest of the chorus frogs, Strecker's Chorus Frog is stouter and paler than its closely related eastern cousin the Ornate Chorus Frog. The species is named in honor of John K. Strecker (1875–1933), a naturalist collector who is considered to be the father of Texas herpetology. Isolated populations in Illinois and Missouri have been recognized by some authorities as a separate species, the Illinois Chorus Frog *(P. illinoensis),* but recent studies suggest that there is only one species.

Appearance: There is a dark stripe through the eye that thickens behind it, and a dark spot or bar of variable size beneath the eye. The forelimbs are noticeably thick, and the fingers of the forefeet are chubby. The dorsal ground color may be brown, gray, olive, or green, and there are often contrasting dark spots on the back. Unlike other chorus frogs in its range, Strecker's Chorus Frog lacks a continuous thin light line along the upper lip.

Range and Habitat: Ranges widely in eastern Texas and Oklahoma, with isolated populations in Arkansas, southeastern Missouri, and Illinois. In the main part of its range, it may be found in a wide variety of habitats, from rocky streams in the hill country in central Texas to moist woodlands, sand prairies, and agricultural fields. In southeastern Missouri and Illinois, however, breeding sites are almost always in areas with sandy soils.

Behavior: Like Ornate Chorus Frogs, males tend to call from more exposed positions than other chorus frogs. Breeding takes place in winter in the southern part of the range and early spring in the north. Burrows when not breeding, using its thick front legs to dig forward instead of backward, similar to the behavior of the true toads and spadefoots.

Voice: The advertisement call is a sharp, metallic *peep* or *pip* that is rapidly repeated: *pip-pip-pip-pip-pip.* It sounds much like the call of the Ornate Chorus Frog (a southeastern species with a nonoverlapping range) but is slightly lower in pitch.

Midland Chorus Frog

Pseudacris triseriata (3/4″– 1 1/2″)

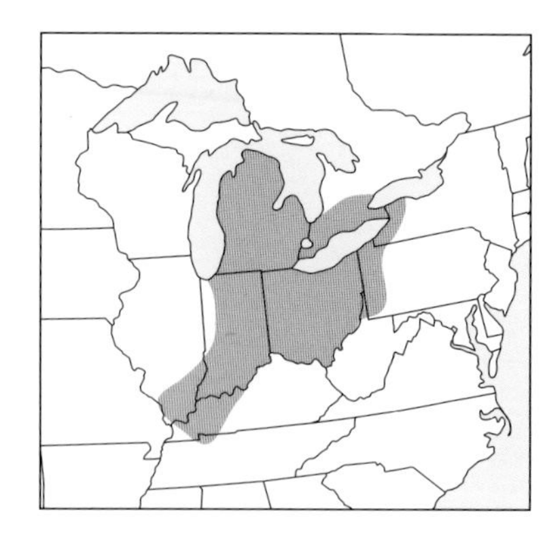

This species was formerly called the Western Chorus Frog and was thought to range well into the West, but recent studies of this frog and several closely related chorus frogs revealed that it had a much smaller range in the East. It has therefore been given a new name, the Midland Chorus Frog. Midland Chorus Frogs are closely related to New Jersey and Upland Chorus Frogs, and all three were previously considered to be subspecies of the Western Chorus Frog.

Appearance: Like most of its close relatives, the Midland Chorus Frog is slender, has a pointed snout, and usually has three longitudinal stripes on the back in addition to dark stripes on both sides of the body. The back stripes, which are sometimes broken, are dark brown or gray and contrast with a lighter gray, brown, or greenish ground color. There is a distinct white line along the upper lip.

Range and Habitat: Found from western New York and Pennsylvania to Michigan, Ohio, and most of Indiana. Also occurs in southern Illinois, western Kentucky, and southeastern Missouri. Breeding sites are temporary pools, ponds, swamps, and roadside ditches in a wide variety of habitats, ranging from prairies to floodplains and moist forests.

Behavior: Midland Chorus Frogs are cool-weather breeders that usually call from dense vegetation. They often adopt a very upright calling posture so that their bodies and stripes are somewhat parallel to blades of grass, making them very difficult to find.

Voice: The advertisement call is a rapid clicking trill lasting a little less than a second and rising slightly in pitch: *crrreeeeek!* The sound can be approximated by running a thumb or finger along the teeth of a pocket comb. A very similar call pattern occurs in the Boreal and Upland Chorus Frogs, which border its range to the west and south.

Upland Chorus Frog

Pseudacris feriarum (3/4″– 1 1/2″)

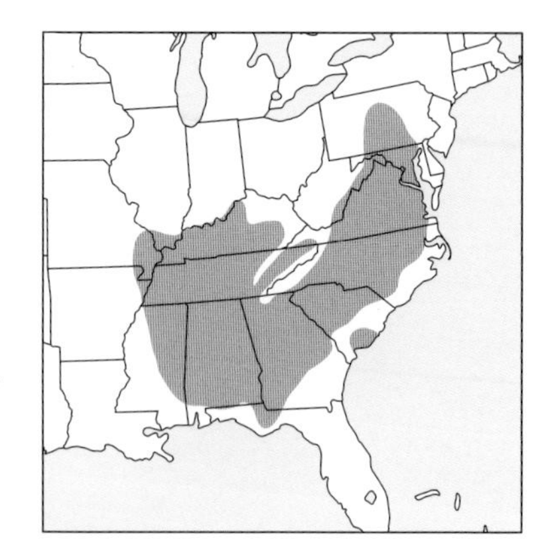

Closely related to the Midland and New Jersey Chorus Frogs and similar in appearance and sound, the Upland Chorus Frog can usually be identified by its geographic location, except in areas of overlap with the Southern Chorus Frog, along the southern edge of its range. This frog was given the name *Upland* because it ranges throughout most of the piedmont region of the United States, whose rolling hills are higher in elevation than the coastal plain habitat of the Upland Chorus Frog's southern relative.

Appearance: The ground color varies from dark brown to light tan. Darker markings on the back may appear as straight longitudinal stripes, ragged and broken stripes, or spots of various sizes. Some individuals have no pattern. As in its relative the Midland Chorus Frog, there is a dark lateral band from the front of each nostril to the groin, and a white line runs along the upper lip.

Range and Habitat: Found mainly in the upper coastal plain and piedmont from central Pennsylvania south to the Carolinas and then west to Mississippi. The range also reaches northward into southeastern Missouri, Tennessee, and Kentucky. The species is mostly absent from the Appalachian Mountains. An isolated population occurs in the lower coastal plain of South Carolina, and there is a fingerlike extension of the range southward into the Florida Panhandle. This frog breeds in a wide variety of temporary, fishless bodies of water. The nonbreeding habitat includes marshes, river floodplains, wet woodlands, and areas near ponds and bogs.

Behavior: Upland Chorus Frogs are typically early breeders and may be heard calling from December to May, depending on the local climate. They may also call after heavy rains in the summer. Calling males are cryptically colored, assume calling positions that enhance their camouflage, and usually stop calling at the least disturbance.

Voice: The advertisement call is a clicking trill with a rising inflection: *crrreeeeek!* It is difficult to distinguish from the calls of Boreal and Midland Chorus Frogs but has a noticeably faster pulse rate (within trills) than Southern and Cajun Chorus Frogs' calls.

New Jersey Chorus Frog

Pseudacris kalmi (3/4″– 1 1/2″)

Restricted in range to southern New Jersey and surrounding areas, the New Jersey Chorus Frog was formerly considered to be a subspecies of the Upland Chorus Frog, but it is somewhat larger and stouter and may have a slightly different call. However, the two species are best distinguished by their different geographic ranges.

Appearance: The ground color varies from gray to brown and sometimes greenish. The three dark back stripes are usually thick and well defined, although in some individuals the stripes are dull or irregular. There is a broad dark band on each side running from in front of the nostril, through the eye, and to the groin. As in other chorus frogs, there is a light line above the upper lip.

Range and Habitat: Found in the southern two-thirds of New Jersey and the Delmarva Peninsula of Delaware, Maryland, and Virginia. Populations also occur on Staten Island, New York, and in extreme southeast Pennsylvania along the New Jersey border. Prefers hardwood and mixed hardwood–pine forests.

Behavior: Breeds from February to April, in temporary and semipermanent bodies of water, including ditches, ponds, streams, and bogs. Males arrive before females and call from grassy clumps. Outside the breeding season, adults are occasionally encountered in wooded areas.

Voice: The advertisement call is a rapid clicking trill: *crrreeeeek!* It sounds almost identical to the call of the Upland Chorus Frog, but usually has a slightly higher pulse rate.

Cajun Chorus Frog

Pseudacris fouquettei (7/8"– 1 1/4")

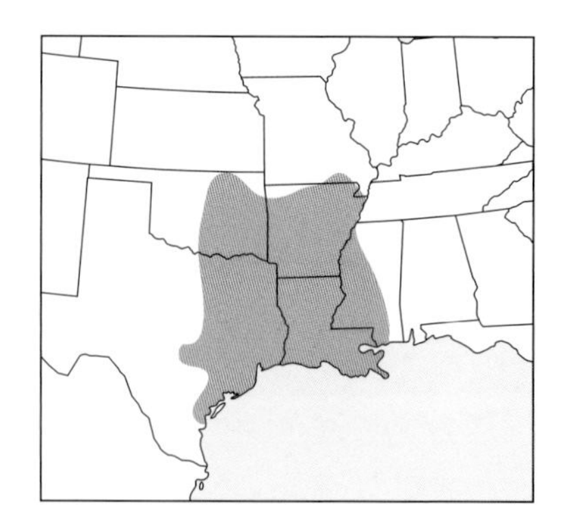

Recently described as a new species, populations of the Cajun Chorus Frog were formerly considered to be either Upland or Midland Chorus Frogs, depending on location. The common name of the species celebrates the fact that its type locality (the location from which the specimen used to define the species was collected) is in Cajun country, in the Baton Rouge area of Louisiana. The scientific name honors M. J. Fouquette, a pioneer in the study of chorus frog vocalizations. Genetic studies show that the Cajun Chorus Frog is most closely related to the Southern Chorus Frog.

Appearance: This species looks almost identical to the Upland and Boreal Chorus Frogs that occur along the edges of its range and is best distinguished by location or call. Ground color ranges from dark brown to tan, and dark stripes or spots occur on the back. A dark stripe extends laterally from the front of each nostril to the groin, and there is a white line along the upper lip.

Range and Habitat: Occurs along the Gulf Coast from western Mississippi to eastern Texas and north to eastern Oklahoma, Arkansas, and extreme southern Missouri. Found in a wide variety of habitats, usually near streams, pools, lakes, and other wet areas.

Behavior: Breeding occurs from winter to spring in temporary and semipermanent ditches, pools, and ponds. Like most chorus frogs, individuals may be found some distance from water outside the breeding season.

Voice: The advertisement call is a clicking trill that usually lasts a little over a second: *crrreeeeeek!* It sounds very similar to the calls of other chorus frogs that produce clicking trills but shares with the Southern Chorus Frog the trait of having relatively long call durations coupled with slow pulse rates within trills; Upland Chorus Frogs to the east and Boreal Chorus Frogs to the north have noticeably faster pulse rates within trills.

Boreal Chorus Frog

Pseudacris maculata (3/4"–1 7/16")

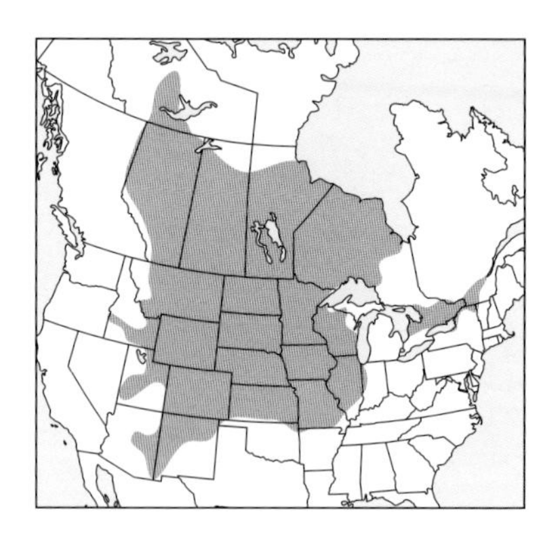

Boreal Chorus Frogs have by far the largest range of any of the chorus frogs. They get their common name from their prevalence in the boreal forests of the northern Midwest and Canada, but recent studies have shown that their range is actually much larger than originally thought, while the range of their cousin the Midland Chorus Frog (*P. triseriata*—formerly called the Western Chorus Frog) is much smaller.

Appearance: The lower legs of Boreal Chorus Frogs are proportionally shorter than those of other chorus frogs, a difference that may be an adaptation to cold climates. The back usually shows three continuous dark stripes or broken dark spots in three rows.

Range and Habitat: The range is vast, extending from Missouri westward to Colorado and New Mexico, and then northward into Canada to the Hudson Bay, the Northwest Territory, and the Yukon. Along the western edge of the range, there are fingerlike extensions into Idaho and extreme eastern Oregon, Utah, and Arizona. The Boreal Chorus Frog overlaps along the eastern and southeastern edge of its range with Midland, Upland, and Cajun Chorus Frogs. Hybridization is likely in these areas, where species identification is difficult if not impossible without assessing genetic data. Boreal Chorus Frogs are found in and around marshy areas, lakes, pools, and a variety of other semipermanent bodies of water.

Behavior: Breeding occurs from February to June, depending on location. Calling is usually initiated by rainfall, and the frogs use both semipermanent and temporary breeding sites. Because of their relatively short legs, Boreal Chorus Frogs try to escape from predators by hopping rather than by making long leaps, as other chorus frogs do.

Voice: The advertisement call, which lasts a little over a second, is a rapid series of metallic clicks that rise in pitch: *crrreeeeek!* Calls are typically repeated one after the other, with two or three seconds of silence in between. They sound very similar to the clicking trills of Midland and Upland Chorus Frogs but have noticeably faster pulse rates within trills than Cajun Chorus Frogs (all these species occur along the southeastern corner of the Boreal Chorus Frog's range).

Southern Chorus Frog

Pseudacris nigrita (3/4"– 1 1/4")

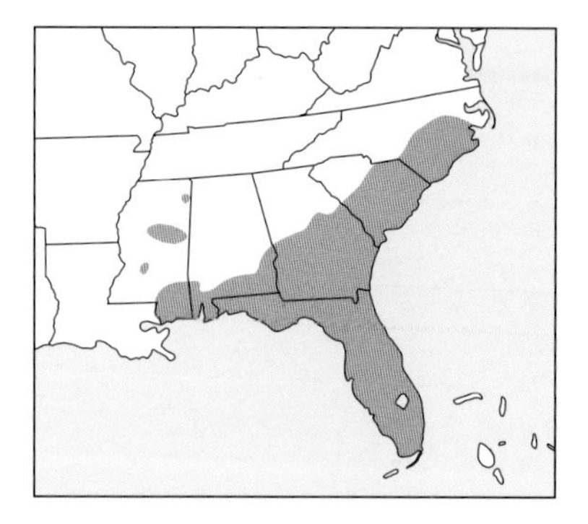

True to its name, the Southern Chorus Frog ranges farther south than any other members of its genus, reaching the southern tip of mainland Florida. A denizen of pine and mixed woodlands, its telltale creaking call may be heard during warm spells from midwinter into spring.

Appearance: The Southern Chorus Frog often looks darker than our other chorus frogs because of the large irregular black markings on its back, commonly arranged as three broken stripes. The ground color ranges from tan or light gray to brown. There is usually a prominent white line along the upper lip. Individuals from coastal plain areas show a strong tendency toward striping on the back, while those in the Florida peninsula are more likely to be heavily spotted on the back, with the white upper lip interrupted by black. Throughout the range, individuals that exhibit little striping or spotting are occasionally found.

Range and Habitat: Found in the coastal plain region from southeastern Virginia to Mississippi, including peninsular Florida. Prefers pine flat-woods and other relatively well-drained and acidic habitats. Inhabits limestone sinkholes bordering wet prairies in southern Florida.

Behavior: The breeding season is prolonged in many areas, beginning as early as December or January and extending into April or May. The situation is different in southern Florida, where breeding usually peaks in June and July with the arrival of heavy rains. Males call from flooded grassy areas, pools, ditches, and ponds. They are secretive and cryptically colored, hide in grass clumps and other vegetation (sometimes with only their heads protruding from the water), and usually stop calling at the least disturbance.

Voice: The advertisement call is a slow-paced clicking trill that rises in pitch and lasts about a second: *crrreeeeek!* It sounds much like the slow-paced trill of the Cajun Chorus Frog and has a noticeably slower pulse rate (within trills) than that of the Upland Chorus Frog, with which it overlaps in range.

Brimley's Chorus Frog

Pseudacris brimleyi (1"– 1¼")

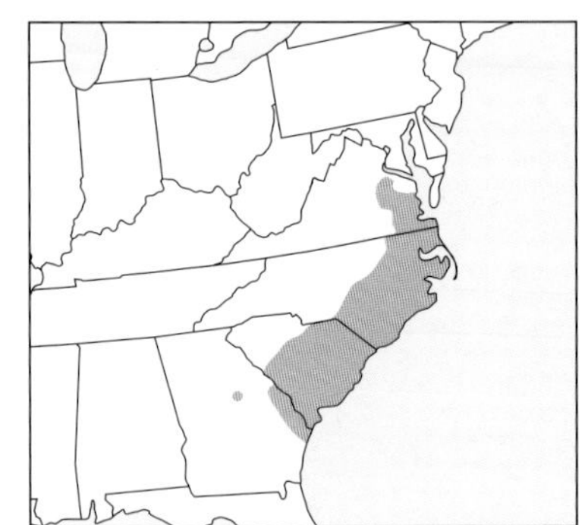

The slender and streamlined Brimley's Chorus Frog is named in honor of C. S. Brimley, a noted North Carolina herpetologist of the early 1900s. Restricted mostly to coastal plain habitats in the Carolinas, it is often heard giving its raspy calls from ditches and pools during warm spells in the middle of the winter.

Appearance: The dorsal color ranges from various shades of brown to light gray, but the typical individual has a straw-colored back that contrasts with three somewhat darker longitudinal stripes. In addition, there are prominent black stripes that extend from the front of each nostril and back along the side to the groin, and there is a distinct light line along the upper lip. While the back stripes may be faded and barely visible in some individuals, the side stripes are always bold and obvious. The easiest way to distinguish this species from other chorus frogs is by the dark longitudinal bars or stripes on the rear legs—other chorus frogs have transverse bars.

Range and Habitat: Found in the Atlantic coastal plain and piedmont regions from Virginia to Georgia. Inhabits hardwood forests and swamps, but may also be found in drier habitats such as pinewoods and agricultural fields.

Behavior: Brimley's Chorus Frogs are cool-weather breeders and do most of their calling from December through March. Males gather in temporary pools, ponds, roadside ditches, and swampy areas that are in or near the floodplain of a river or stream. The well-hidden males usually adopt an upright calling posture in clumps of thick grass or other vegetation, making them exceedingly difficult to find.

Voice: The advertisement call is a raspy trill, repeated up to two times per second: *rrrack-rrrack-rrrack-rrrack-rrrack*. It might be confused with the raspy call of the Squirrel Treefrog, which may be found in the same locations, but Squirrel Treefrogs typically do not breed until much later in the spring and summer.

Mountain Chorus Frog

Pseudacris brachyphona (1″– 1½″)

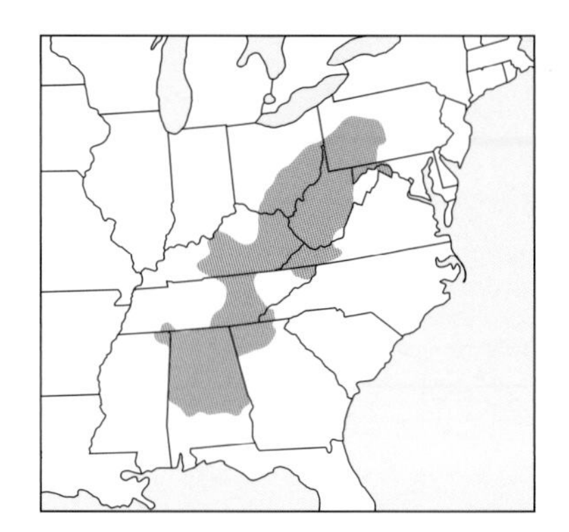

The Mountain Chorus Frog is a small, stocky woodland species found in the mountains and foothills of the Appalachians. It is easily mistaken for a Spring Peeper but has smaller toe pads. Although occasionally observed hopping about in moist woodlands, the Mountain Chorus Frog is perhaps best detected in spring by listening for its raspy trill, which is quite unlike the calls of other frogs in its surroundings.

Appearance: The dorsal ground color ranges from brown or gray-brown to olive, and there is yellow in the hidden parts of the legs. There is a dark mask through the eyes and sometimes a dark triangle between the eyes. As in other chorus frogs, there is a white line or band along the edge of the upper lip. An average individual has dark crescents on the back that bow inward like reverse parentheses, sometimes forming an × that resembles the pattern on the back of a Spring Peeper (but peepers generally lack the white line on the lip). Still others may exhibit a variously broken pattern of three stripes down the back, resembling the stripes of Midland and Upland Chorus Frogs. Plain-colored individuals with no obvious back pattern are common at the southern end of the range.

Range and Habitat: Found in the Appalachian Mountains and associated foothills from Pennsylvania south to Alabama. Prefers forested hilly terrain and avoids mountain valleys and other flat areas where other chorus frogs might be found.

Behavior: Mountain Chorus Frogs are early-spring breeders and may be heard calling from February into April. They breed in a wide variety of shallow and usually temporary bodies of water, including roadside ditches, flooded areas, small streams, and ponds or pools around springs.

Voice: The advertisement call is a harsh, raspy trill with an up-slurred ending that is repeated around two times per second: *rrrack-rrrack-rrrack-rrrack-rrrack.* It sounds very similar to the call of its closest relative, Brimley's Chorus Frog. The ranges of the two species do not overlap.

Spotted Chorus Frog

Pseudacris clarkii (3/4″– 1 1/4″)

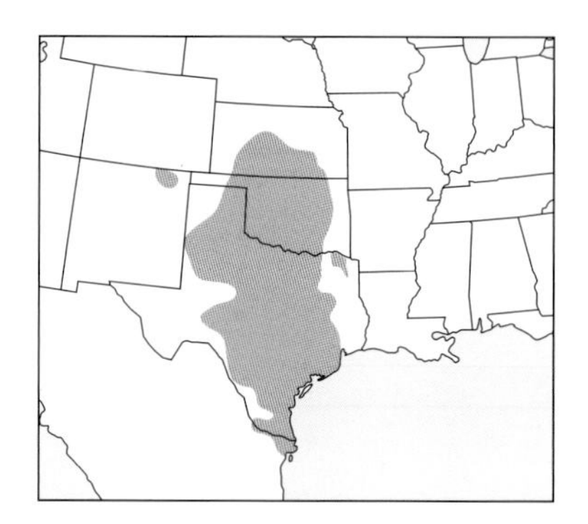

A prominent pattern of spots justifies the common name of the Spotted Chorus Frog, which is among the most colorful of its group. Genetic studies have revealed a history of hybridization with the Boreal Chorus Frog, but the two species look different, have distinctly different calls, and rarely interbreed where their ranges overlap (in south-central Kansas).

Appearance: The dorsal ground color is usually some shade of pale gray, olive, or green, and the spots are typically dark green with a black border. Some individuals are very ornamented, with rich green spots set against a light tan background, or dark green spots arranged over lighter patches of sparkling green. There is usually a triangular-shaped green area between the eyes. In some individuals, the spots can be fused to give the look of stripes running up and down the back. Other identifying features include dark bands that run from each nostril to the midsection or groin and a light stripe along the upper lip.

Range and Habitat: A prairie grassland species that ranges from Kansas to southern Texas (and into Mexico). Also occurs in agricultural and other human-disturbed open areas that were once prairie lands.

Behavior: Spotted Chorus Frogs are explosive breeders that call from temporary ponds, pools, ditches, and other flooded areas after heavy rains. Breeding occurs in most areas in early spring, usually from February to April, following the first warm rains of the season. In the southern part of the range, breeding may occur almost anytime during the year, depending on rainfall. Males call from grassy edges of wetlands, where they remain immersed, only their heads and vocal sacs exposed. Outside the breeding season they wander far from water.

Voice: The advertisement call is a rapidly repeated, scraping trill, *rrrack-rrrack-rrrack-rrrack-rrrack,* that sounds very similar to the calls of the more easterly Mountain and Brimley's Chorus Frogs. May also be confused with the call of the Squirrel Treefrog, which overlaps in range with the Spotted Chorus Frog in eastern Texas.

Pacific Chorus Frog

Pseudacris regilla (3/4″–2″)

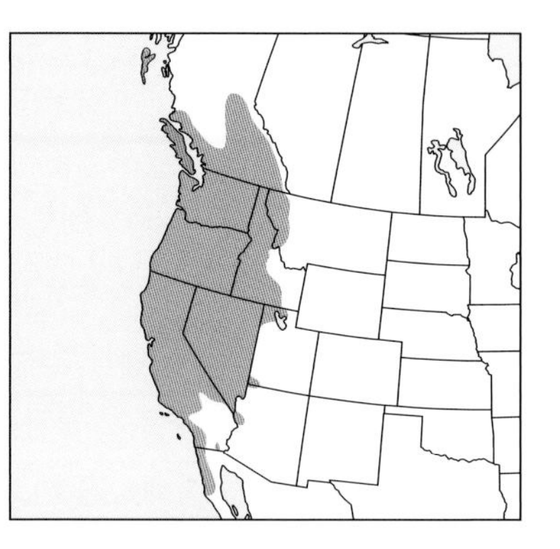

The small, variably colored Pacific Chorus Frog is the most commonly heard frog in the Pacific coastal states. Its familiar two-part call, a croaking *rib-bit,* is the sound that the general public has come to associate with frogs, mainly because it is frequently heard in the background of Hollywood films. Possessing fairly well-developed toe pads, the species was once included in the genus *Hyla* and referred to as the Pacific Treefrog, but molecular data suggest a closer affinity with the chorus frogs. Recent mitochondrial studies indicate that there may be three separate species, geographically arranged from north to south. The proposed new names are Northern Pacific Chorus Frog *(P. regilla),* Sierra Chorus Frog *(P. sierrae),* and Baja California Chorus Frog *(P. hypochondriaca).*

Appearance: Aside from a black or brownish eye stripe, which is almost always present, the coloration of the Pacific Chorus Frog is highly variable. Ground color can be green, brown, beige, copper, tan, rust, or even black, and various stripes or spots in an equally varied number of colors may be present on the back and legs. (The accompanying photographs depict a number of variations.) Individuals can change color in just a few minutes.

Range and Habitat: Ranges from Baja California (Mexico) north into British Columbia and eastward to the Rocky Mountains. Breeds in a wide variety of temporary and permanent bodies of water. When not breeding, it is typically found on the ground and in low vegetation, often near water.

Behavior: Pacific Chorus Frogs are cool-weather breeders, but because they occur from sea level to high in the mountains, they can be found calling somewhere in their range from November to July. Males use advertisement calls to space themselves and switch to aggressive calls when rivals get too close. Calling occurs mainly in the evening and at night, although males may vocalize sporadically during the day at the height of the breeding season.

Voice: The advertisement call usually has two distinct parts, with the second part rising slightly in pitch: *rib-bit, rib-bit* or *kreck-ek, kreck-ek.* Geographic variation exists in this call, with some being single-parted. The aggressive call is a drawn-out creaking trill, *kr-r-r-r-r-r-eck.* Away from breeding ponds, males may give a single-note *krrreck* that sounds similar to the advertisement call of the California Chorus Frog but is repeated more slowly.

California Chorus Frog

Pseudacris cadaverina (1″–2″)

California Chorus Frogs (also called California Treefrogs) look much like the rocks on which they are found. Their scientific name, *cadaverina,* is Latin for "having the appearance of a corpse" and refers to the pale, corpselike coloration of the species. California Chorus Frogs look and behave like Canyon Treefrogs, but they are smaller and have distinctly different calls, and genetic studies show they are more closely related to Pacific Chorus Frogs.

Appearance: California Chorus Frogs have well-developed toe pads and webbing and a dorsal ground color of light gray or brown with darker, irregular blotches. The skin of the upper body has small wartlike bumps. The underside is white, and there is some yellow on the hidden parts of the hind legs.

Range and Habitat: Restricted in range to arid lowlands, foothills, and mountains of southwestern California and the northern Baja Peninsula of Mexico, where the preferred habitat is a deeply cut canyon dominated by granite rocks and permanent pools. They are most often found in canyons with east- and west-facing slopes and may occur from sea level up to about 7,500 feet.

Behavior: California Chorus Frogs breed in rocky streams, where they occur year-round. Most mating takes place from February through October. Males usually call from the shoreline and from rocks in and near the water. During the day, individuals often hide under rocks and in crevices but may also be found sunning on rocks, where they are hard to see because of their cryptic coloration. When exposed, they secrete a clear fluid that protects their skin from drying. If disturbed, they will leap into the water and then swim quickly to shore.

Voice: The advertisement call is an explosive quacklike note that is repeated once every second or so: *wreck! wreck! wreck! wreck!* It sounds similar to the one-part advertisement calls of certain populations of Pacific Chorus Frogs. During aggressive encounters, a more raspy, scraping note is given that sounds much like the dry-land call of the Pacific Chorus Frog.

Northern Cricket Frog

Acris crepitans (5/8″– 13/8″)

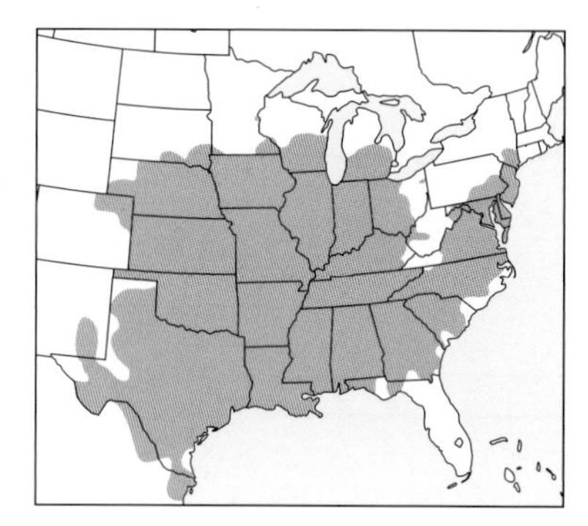

Although a member of the treefrog family, the Northern Cricket Frog lacks toe pads and seldom climbs. Scarcely longer than an inch, these small frogs are wary and difficult to catch and can leap long distances for their small size.

Appearance: Adults have warty skin and highly variable coloration. Ground color can be brown, green, or gray. There is usually a dark triangle between their eyes. Most individuals have prominent dorsal stripes that are green, yellow, orange, or red. (The accompanying photographs show variations.) Large spots occur along their sides, and the top part of their thighs have narrow, dark bands. Also, a dark, ragged stripe runs lengthwise along the rear edge of the thigh that contrasts with a light ground color (this line is more clear-cut in the similar-looking Southern Cricket Frog). In the western part of the range, the Blanchard's Cricket Frog *(A. crepitans blanchardii)*, which may soon be elevated to a full species, is larger and chunkier and tends to be more uniformly colored than frogs in eastern areas (pictured in top photo below).

Range and Habitat: Wide-ranging in the East, although notably absent from peninsular Florida and northern and mountainous areas. Commonly found along the edges of lakes and ponds. In the western part of the range, they are likely to be found in very shallow water in small streams and ditches. Populations in Florida and the lower coastal plains are restricted to bottomlands along major rivers.

Behavior: The breeding season extends from late winter well into summer. Although most calling occurs at night, cricket frogs often call for short periods during the day, especially early in the breeding season when temperatures are cool.

Voice: The advertisement call consists of a series of clicks that sound like two small stones being tapped together: *gik-gik-gik-gik.* Clicks are repeated slowly at first and then organized into pairs or triplets that build into a final extended terminal series of excited calls, then slow down at the end. Individual clicks *(giks)* are more sharply defined than the more drawn-out clicks *(giiks)* of the Southern Cricket Frog.

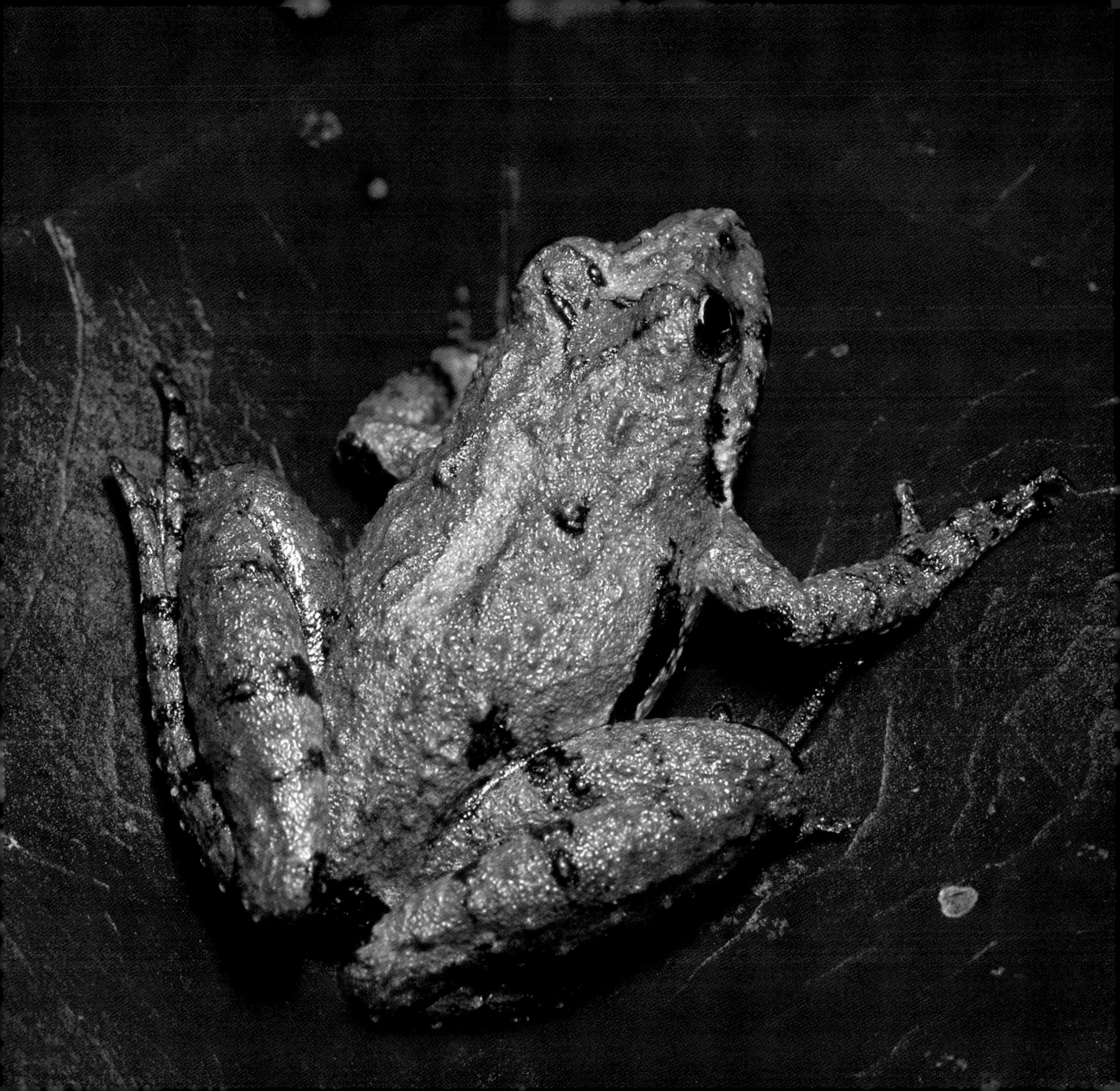

Southern Cricket Frog

Acris gryllus (5/8"– 1 1/4")

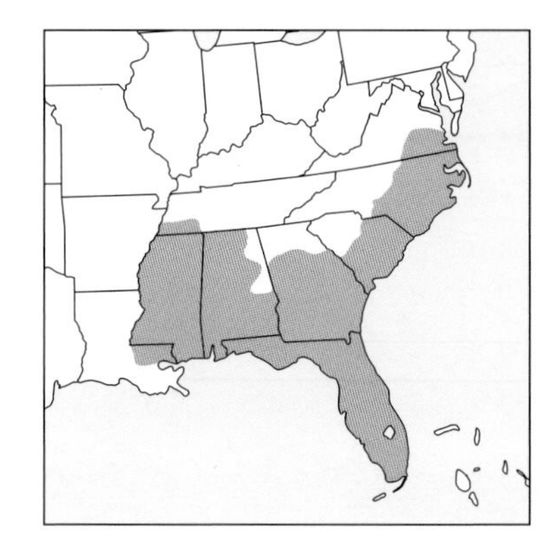

Because they are so similar in size, appearance, and sound, Southern and Northern Cricket Frogs were once considered subspecies of a single species. When attempting to escape, individuals of both species make rapid long leaps, often jumping quickly into the water and then out. This behavior makes them much more likely to be seen than chorus frogs and other hylids, which usually rely on camouflage and staying still to avoid detection.

Appearance: The two species of cricket frogs can be difficult to tell apart. They both show remarkable color variation, especially with respect to their mid-dorsal stripes, which range from gray to vibrant green or red. Yet they can be distinguished both by their body shapes and calls. The Southern Cricket Frog typically is more slender and long-legged than its northern counterpart, and it has a more pointed snout. It also has less webbing on its hind feet, and an outstretched hind leg has a well-defined dark stripe (or stripes) running lengthwise on the rear of the thigh (frogs must be captured for this to be seen). In Northern Cricket Frogs these stripes are much less distinct and have a ragged appearance.

Range and Habitat: A lowland frog of the southeastern coastal plain, ranging from eastern Louisiana to Virginia. Found in a variety of moist habitats, including swamps, bogs, lake edges, and ponds.

Behavior: May be heard any month of the year, but breeds primarily from April through summer. Males call from floating vegetation and from the shorelines of lakes, ponds, pools, and streams.

Voice: The advertisement call is a metallic click or *giik* given in groups of one to ten or more and delivered at a fairly even pace: *giik, giik, giik, giik.* An individual typically starts with one *giik* and then adds one or two *giik*s to each subsequent series, as if counting. The clicks of the Southern Cricket Frog are audibly more drawn out than those of the Northern Cricket Frog, and the terminal series of clicks is not as rapid-fire and doesn't wind down at the end as in the Northern Cricket Frog's.

Cuban Treefrog

Osteopilus septentrionalis (1½″–5½″)

Reaching lengths of five inches or more, this impressive invader from the West Indies continues to expand its range in Florida, where it eats a variety of creatures, including smaller native frogs and toads—a common negative effect of invasive species. Cuban Treefrogs frequently invade homes, coming in through open doors and windows and then hiding in all manner of places, even inside toilet bowls.

Appearance: The adult can be identified by its large size, broad head with the skin tightly fused to the skull, extra-large toe discs, and rough skin. A smaller individual can be distinguished from other Florida treefrogs by its lack of stripes and other markings running from the front of the eye to the rear and sides of the body. Individuals can change color rapidly, usually from some shade of brown or tan to pale green (or vice versa), and spots may appear or disappear.

Range and Habitat: Found throughout southern Florida and expanding its range northward (isolated individuals have been found as far north as South Carolina, and this species has also become established on Oahu Island in Hawaii). Cuban Treefrogs spread as stowaways on ornamental plants, boats, motorized vehicles, and the like. They frequent residential areas, where they spend the day in moist, shady places in and near homes and outbuildings, often revealing their presence with raspy, barking calls. Natural retreats include trees, shrubs, and palm fronds.

Behavior: Cuban Treefrogs are excellent climbers and usually sleep aboveground during the day. At night they forage near lights that attract insects. They attempt to eat anything that's moving and is smaller than they are, including snails, spiders, insects, frogs, and even small lizards and snakes. Breeding occurs from March to October and is stimulated by heavy rainfall, especially warm summer rains associated with tropical storms. They breed in almost any wetland that is free of predatory fish, including ponds, pools, ditches, and even swimming pools!

Voice: The advertisement call is a grating, nasal *raaaack,* repeated about once a second. Males may call from trees and other hideouts away from breeding areas, especially in response to rain showers.

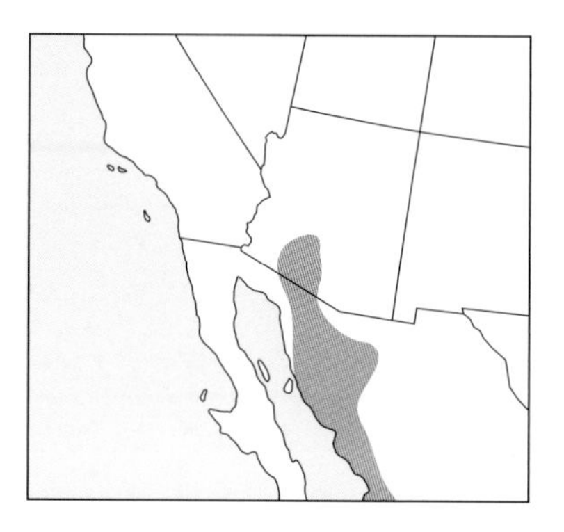

Lowland Burrowing Treefrog

Smilisca fodiens (1½″–2½″)

The Lowland Burrowing Treefrog is a unique-looking member of the treefrog family that is adapted for a fossorial, or burrowing, existence. It is also called the Northern Casque-headed Frog because the skin on its head is tightly fused to the skull, making it resemble a casque or helmet. First discovered in the United States in 1957, the species barely ranges into southern Arizona and is seldom seen except when breeding after summer monsoon rains.

Appearance: Stout in build, this frog has legs that are short and stocky, and each hind foot has a tough spadelike tubercle that aids in digging. There is also a characteristic fold of skin at the back of the head. The usual ground color is light brown or pale tan with distinct, black-bordered dark spots. (Juveniles are green.) Toe pads are evident but are not nearly as large as those of many other treefrogs. The vocal sac is slightly bilobed.

Range and Habitat: Although common and widespread in parts of Mexico, the Lowland Burrowing Treefrog has a very limited range in the United States. It is found only in the Sonoran Desert of extreme southern Arizona. Most of its range lies within the territory of the nation of the Tohono O'odham ("People of the Desert"), where it is found mainly in valley bottoms that accumulate rainwater, often at sites grown over with mesquite.

Behavior: Lowland Burrowing Treefrogs breed during the early monsoon period, when summer rains fill dry arroyos, roadside ditches, and cattle wallows. Nighttime choruses may be dense and are often audible from long distances. Males frequently call from secluded locations, often out of water but near the breeding sites. Outside the breeding season, individuals remain underground in self-made burrows and abandoned rodent burrows. They use their heads to seal their burrow entrance and may shed several layers of skin to form a protective cocoonlike covering that helps them retain water during the dry season.

Voice: The advertisement call is a repeated ducklike *wauk, wauk, wauk, wauk* that sounds like a low-pitched version of the call of the Arizona Treefrog. The calls of the two species cannot be confused in the wild, however, because their geographic ranges do not overlap.

Mexican Treefrog

Smilisca baudinii (2″–3½″)

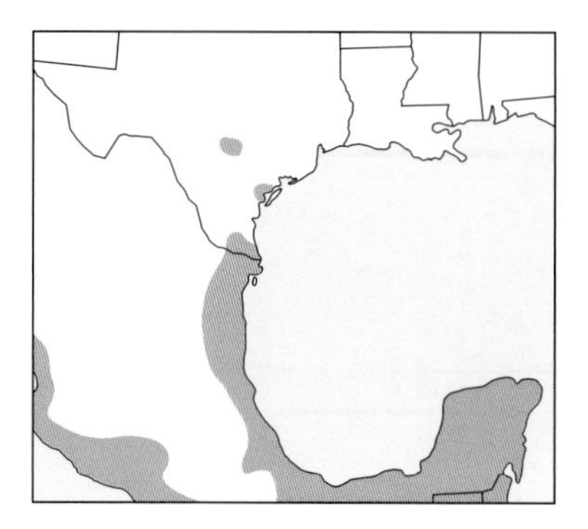

The Mexican Treefrog is a beautiful tropical species ranging into the United States only in extreme southeastern Texas, where it is an uncommon resident. It is our largest native treefrog, reaching lengths of three inches or more (the introduced Cuban Treefrog may grow to five inches). Individuals remain inactive during dry periods and can be extremely difficult to find.

Appearance: Mexican Treefrogs have smooth skin and well-developed toe pads. There is a characteristic white spot under the eye, and a dark thick stripe extends from the snout to the shoulder. The body color and pattern are extremely variable, and individuals are chameleon-like, able to change color quickly from tan to brown or green. The vocal sac is distinctly bilobed.

Range and Habitat: Found primarily in Mexico, but ranges south to Costa Rica and north to the Rio Grande Valley in Texas, where it is uncommon and occurs only in counties near the mouth of the river. Isolated records farther north in Texas may represent accidental introductions. Found in forested or brushy areas near streams and resacas (oxbow lakes) and in suburban areas where lawns are watered regularly.

Behavior: Mexican Treefrogs breed explosively at any time during the year after heavy rainfall, especially following hurricanes or tropical storms in late summer and autumn. Breeds in nearly any body of water, but prefers small, temporary pools. It may skip breeding entirely during years when big rains do not occur. This species is nocturnal and arboreal, but individuals may take shelter in burrows when it is dry and hide under the bark of dead trees during the day.

Voice: The advertisement call is a rapidly repeated, nasal, honking *heck-heck-heck-heck,* occasionally interspersed with chuckling notes. Two males often alternate their calls in a lively duet.

True Toads — Family Bufonidae

About 480 species belong to the family Bufonidae, a large and diverse group of 33 genera with representatives in nearly every part of the world. They are commonly referred to as true toads, and more than half of these species currently belong to the genus *Bufo* (but see last paragraph), of which 22 species reside in the United States and Canada. This family includes all our familiar terrestrial hop toads.

True toads are typically squat and hefty and they have relatively short legs in comparison to other frogs and toads. They have rough skin and large parotoid glands on their heads. These glands, along with numerous smaller glands found over the entire body, produce noxious and sometimes poisonous secretions called bufotoxins. Most true toads have bony skulls, and many have distinctive ridges on their heads, called cranial crests, which can be very useful in species identification (see diagram below). There is considerable variation in size within the group, from just over an inch long for the tiny Oak Toad to nearly seven inches long for the Cane Toad.

The majority of true toads are terrestrial except during the breeding season, and most are able to burrow in sand or soft soil. They are more likely to hop or crawl than leap. Most species have well-developed advertise-

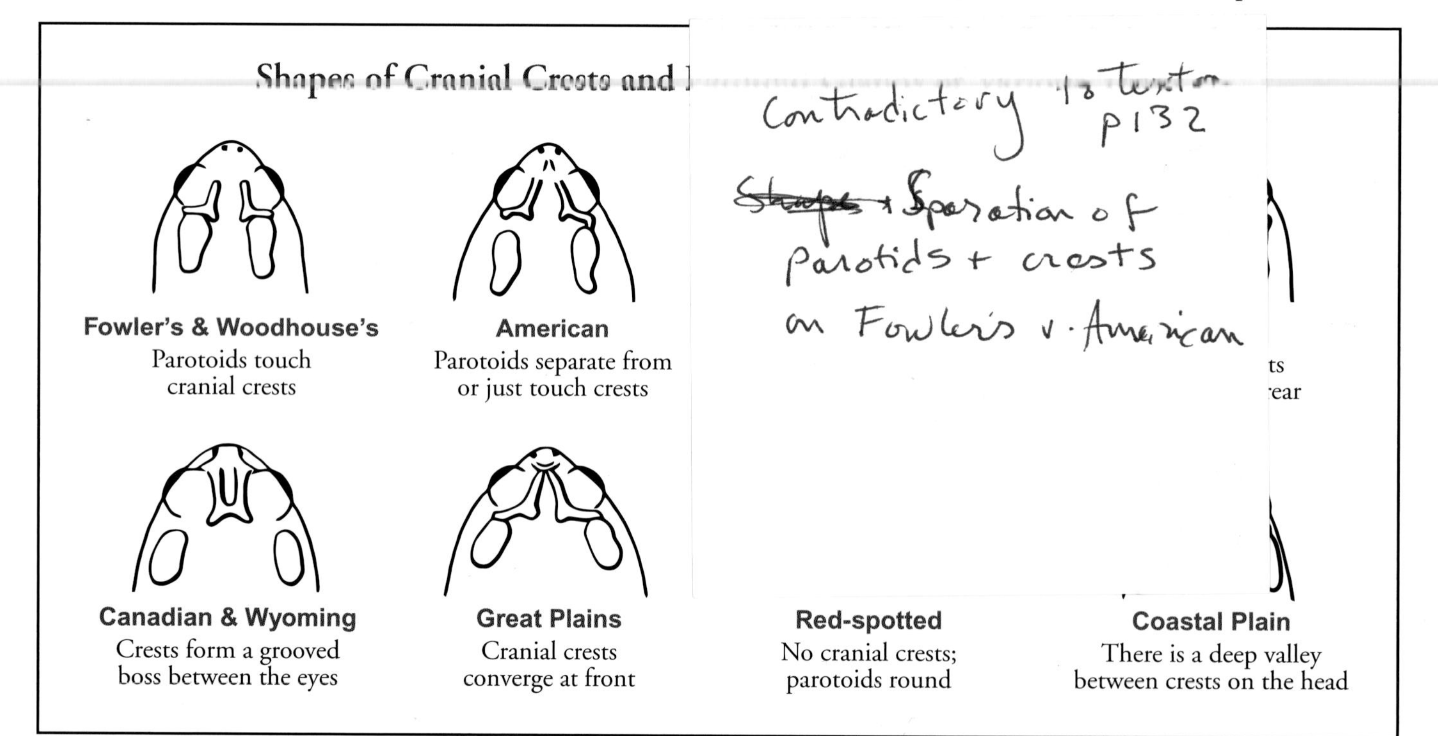

Illustrations from the *Peterson Field Guide to Reptiles and Amphibians of Eastern and Central North America* by Roger Conant and Joseph T. Collins © 1998 by Houghton Mifflin Company

American Toad

ment calls that are often longer in duration than the calls of frogs and toads in other families. Some species, such as the American Toad, give long musical trills. Others, such as the Great Plains Toad and Fowler's Toad, produce harsh or dissonant trills. The tiny Oak Toad doesn't trill at all—it makes high-pitched birdlike peeps.

The relationship among true toads in North America is the focus of considerable debate. In fact, a recent taxonomic study proposes that all our species should be reassigned to new genera. If these changes are accepted by the scientific community, nearly all our toads will be switched from *Bufo* to the new genus *Anaxyrus,* but they will retain their original common names and specific names (for example, the American Toad will be changed from *Bufo americanus* to *Anaxyrus americanus* but will still be referred to as the American Toad). There will also be a few exceptions: the Coastal Plain Toad and Sonoran Desert Toad will be assigned to the genus *Ollotis,* and the Cane Toad to the genus *Rhinella.* Anticipating these changes, we include the new scientific names in parentheses in the headings of our species accounts.

American Toad

Bufo americanus (Anaxyrus americanus) (2″–4⅜″)

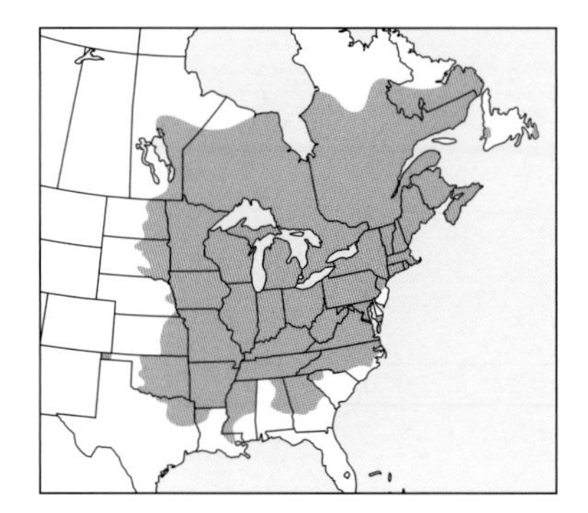

Our most familiar eastern hop toad, the American Toad wanders far from water after the breeding season and is a familiar denizen of yard and garden as well as woodland settings. The beautiful melodic trill of the male is a pleasure to hear—Henry David Thoreau referred to it as "the dream of the toad" in his journal entry of October 26, 1853.

Appearance: Like other true toads, the American Toad has a squat, chunky body and dry, warty skin. Dorsal ground color may be brown, brick red, gray, or olive, and there may be patches of lighter colors. Dark spots are usually evident, and warts (tubercles) are variously colored. There is often a light stripe down the center of the back. In the southwestern corner of the range, Dwarf American Toads *(B. americanus charlesmithi)* are smaller (1½″–2½″), usually reddish in color, and have higher-pitched calls. American Toads are easily confused with Fowler's Toads, but the two species can usually be differentiated by examining the configuration of their cranial crests (page 126).

Range and Habitat: Found throughout the East, including much of eastern Canada, but notably absent from the southern coastal plain. Frequents a wide range of habitats, including residential areas, where they may be found catching insects attracted to outdoor lights.

Behavior: American Toads breed in temporary bodies of water. The breeding season lasts only a week or two, beginning as early as March in southern areas and as late as June at high latitudes and elevations. Males may call from stationary positions and wait for females to come to them or, especially in dense choruses, may move around in search of mates. Fights over females are common, and large males usually displace small ones. American Toads occasionally hybridize with Fowler's Toads.

Voice: The advertisement call is a long musical trill lasting from five to thirty seconds. Each male in a chorus calls at a slightly different pitch, and individuals often alternate and overlap calls in a pleasing manner. When mounted, males and unreceptive females give release chirps accompanied by abdominal vibrations.

An adult Dwarf American Toad

Fowler's Toad

Bufo fowleri (Anaxyrus fowleri) (2"–3 3/4")

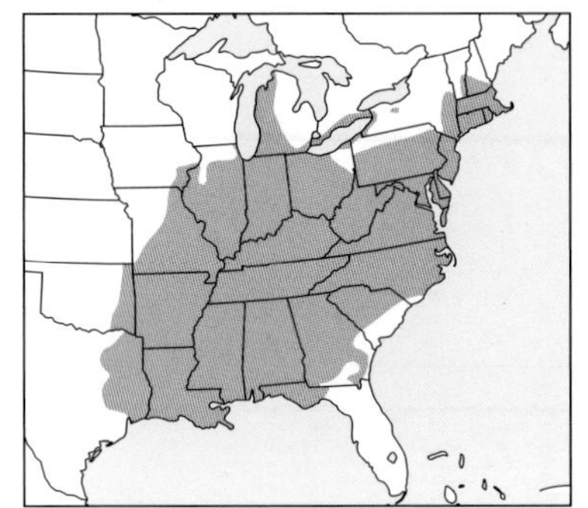

Fowler's Toad is named in honor of Samuel Page Fowler (1800–1888), a New England naturalist who first described the toad in 1843. Formerly considered a subspecies of Woodhouse's Toad, Fowler's Toad ranges throughout much of the East and is often confused with the American Toad, with which it hybridizes.

Appearance: Fowler's Toads look much like American Toads but typically have well-defined dark spots with light edges, and each spot usually contains three or more warts or bumps (American Toads usually have only one or two warts in each spot). In addition, the cranial crests do not usually touch the parotoid glands as they do in American Toads (page 126). Along the western edge of the range, Fowler's Toads readily hybridize with Woodhouse's Toads, and in areas of overlap the two species can be very difficult to distinguish. Indeed, hybridization among Fowler's, Woodhouse's, American, and Southern Toads sometimes make it impossible to identify these toads by appearance alone.

Range and Habitat: Found throughout most of the eastern United States, but absent from parts of the Northeast and upper Midwest, as well as the southeastern coastal plain and the Florida peninsula. The western edge of the range is not well defined because of hybridization with Woodhouse's Toads. Occupies a wide range of habitats, from bottomland hardwood forests to dry pinewoods but prefers sandy and gravelly areas with well-drained soils.

Behavior: Breeds mostly from April to May (earlier in southern areas) in lakes, ponds, rivers, ditches, and pools, but may be heard calling into the summer months. Where it shares its range with the American Toad, Fowler's Toad typically breeds one to three weeks later in the season, but the two species commonly interbreed in years when their breeding seasons overlap. Especially in the Midwest, Fowler's and American Toads tend to occupy different habitats.

Male American Toad clasping a female Fowler's Toad

Voice: The advertisement call is a nasal, buzzy trill lasting from one to five seconds: *waaaaaaaaaaa!* Sounds somewhat like a baby crying and has been described as a wailing scream. Fowler's Toads hybridize with American Toads, and the calls of hybrids are usually intermediate in both harshness and duration between the calls of the parent species.

Woodhouse's Toad

Bufo woodhousii (Anaxyrus woodhousii) (1 3/4"– 5")

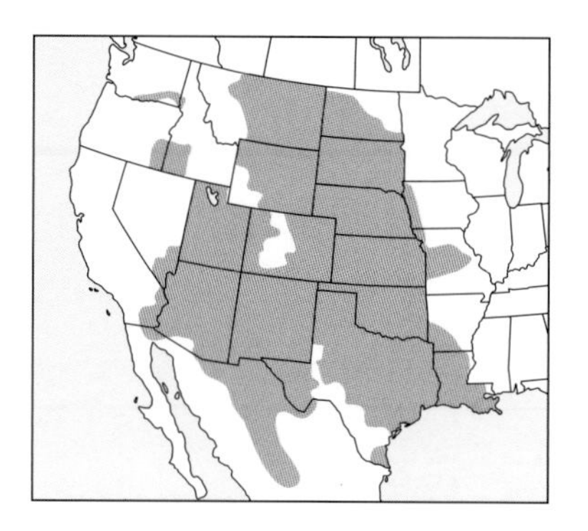

Woodhouse's Toad of the Great Plains and Mountain States is named for Samuel Washington Woodhouse (1821–1904), a naturalist and assistant surgeon who accompanied a U.S. Army expedition to the Southwest in 1851. A relatively large toad, it is highly variable in ground color and pattern of spotting and is often best identified by ruling out all the other toads found in the same area.

Appearance: Woodhouse's Toad is typified by a whitish stripe down the middle of the back, elongate parotoid glands, and prominent cranial crests. Sometimes a slight boss exists between the eyes. Ground color ranges from brown to yellow-brown or gray, sometimes with a hint of olive or green. The underside is light, except in southwestern populations, which have dark markings on the chest. Identification can be especially difficult along the eastern edge of its range, where it hybridizes extensively with Fowler's Toads.

Range and Habitat: Found throughout the Prairie States, from Texas north to near the Canada border, and west to Utah, Nevada, Arizona, and extreme southeastern California. Isolated populations also occur in Idaho, Oregon, and Washington. Frequents diverse open habitats, including farm fields, grasslands, desert scrub, wooded areas, and even city yards and gardens. Most common in and near riparian zones.

Behavior: Breeds from February to August, usually following rains. Breeding males gather wherever water is present: in river bottoms, gulches, ditches, shallow ponds, flooded pools, and lakes. The breeding season is variable in length, lasting about two weeks to two months.

Voice: The advertisement call is a nasal, buzzy *wyaaaaaah;* it is about two to four seconds long, and often begins with a brief up-slurred tone. Sounds similar to the call of Fowler's Toad but is typically lower in pitch (though calls of the two species are more or less indistinguishable in areas where they hybridize).

Southern Toad

Bufo terrestris (Anaxyrus terrestris) (1 5/8″–4 7/16″)

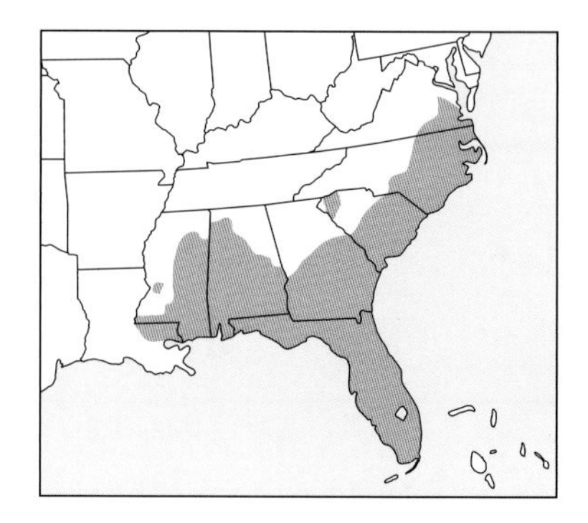

The most common medium-sized toad in parts of the Southeast, the Southern Toad is nocturnal in habits. Individuals spend most of their day in hiding, often in burrows of their own making. At night they come out to feed and may be found around houses and other buildings, where they catch insects attracted by outdoor lights.

Appearance: The Southern Toad is usually some shade of brown, gray, or red, and may have a varied assortment of dorsal spots and warts. Individuals may appear almost black when conditions are cold and wet. Similar in appearance to its more northern cousin the American Toad, the Southern Toad is easily identified by its high cranial crests with pronounced knobs, which are located between and behind the eyes (page 126).

Range and Habitat: Found across much of the South, from Virginia to Louisiana. Primarily restricted to the sandy soils typical of scrub oak, pinewoods, and mixed forests of the coastal plain and peninsular Florida. Often frequents residential areas.

Behavior: Southern Toads are early breeders, with a relatively short breeding season lasting from February to May, during which they may be found in temporary and semipermanent ponds, pools, and ditches. As in other toads, males may call and wait for females to come to them, or they may move around at the breeding site attempting to clasp any other frogs or toads they encounter.

Voice: The advertisement call is a shrill, somewhat melodic trill of about four to eight seconds in duration. It sounds similar to the call of the American Toad but is not nearly as pleasant because of its higher pitch and faster trill rate. Indeed, a group of calling males produces a loud and dissonant chorus that is disagreeable when heard at close range. The release call of the male is a series of grating notes or chirps, accompanied by a soft, low-pitched hum and vibrations of the midsection. Southern Toads hybridize with American Toads, and the calls of hybrids are usually intermediate in character between the calls of the parent species.

Oak Toad

Bufo quercicus (Anaxyrus quercicus) (3/4″– 1 5/16″)

A tiny toad of many colors, the Oak Toad grows to scarcely over an inch in length. It gets both its common name and scientific name from the oak and pine woodlands in which it is most common (*quercicus* is derived from the Latin word *quercinus,* which means "of oak leaves"). In spite of their bright colors, individuals are well camouflaged in their natural surroundings.

Appearance: The ground color varies from light gray to nearly black. There are contrasting dark areas of variable size and shape on the back, arranged symmetrically on each side of a prominent yellowish midline stripe. Small orange-red warts are scattered across the upperparts. The male's vocal sac is large and sausage-shaped, and when the male calls, it curves upward in front of his snout.

Range and Habitat: Found in the coastal plain region from southeastern Virginia to eastern Louisiana and throughout peninsular Florida. There is a northward extension of the range in Alabama. Frequents well-drained sandy pinewoods and pine-oak scrublands, usually where there is little permanent water.

Behavior: Breeds from April to October, but most activity occurs in the spring and summer after warm, heavy rains. Males call from nearly any flooded area, including ditches, pools, puddles, and the edges of ponds. They generally hide in thickets of grass and can be very difficult to find, even when sounding off from a few feet away. Unlike most other toads, the Oak Toad is primarily active during the day, although breeding occurs both night and day.

Voice: The advertisement call is a loud, high-pitched peep, similar to the call of a baby chicken, and briskly repeated, like the sound of a squeaking wheel: *seee-seee-seee-seee-seee-seee.* It is a penetrating sound, and large choruses are overpowering when heard at close range.

Coastal Plain Toad

Bufo nebulifer (Ollotis nebulifer) (2″–5⅛″)

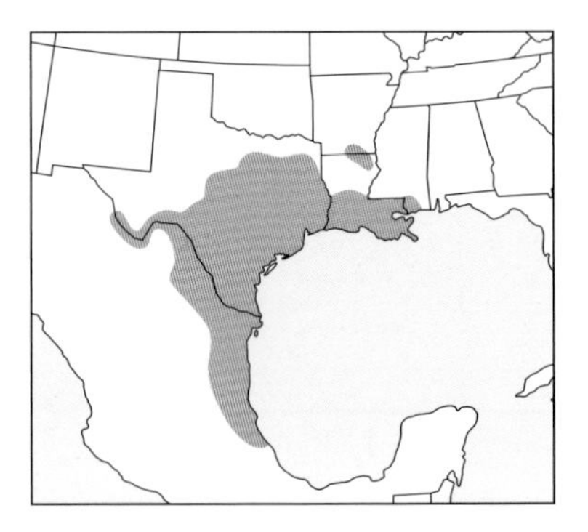

The Coastal Plain Toad—recently recognized as a species distinct from the Gulf Coast Toad *(Bufo valliceps)* that ranges from southern Mexico to Costa Rica—is one of the largest and most commonly seen toads in Texas. Known to be good climbers, Coastal Plain Toads have been found to use tree holes up to fifteen feet off the ground.

Appearance: It can be recognized by its somewhat flattened or squat appearance, along with the distinct concave depressed area between its well-defined cranial crests (page 126). There is a prominent light stripe running down the middle of the back and wide light stripes along each side. The ground color is some shade of brown, ranging from yellow-brown or red-brown to very dark brown, with highlights of white, orange, or gold.

Range and Habitat: Found mostly in the coastal plain region of the Gulf Coast from southern Mississippi to northern Mexico. Common and widespread in Texas, where it ranges westward into the Edwards Plateau and along the Rio Grande River to the Big Bend region. Although found in a variety of habitats, Coastal Plain Toads are not well adapted to arid areas (such as pinewoods and thorn scrub) and are usually found near water. They frequent hardwood forests and agricultural areas as well as suburban and urban environments. They are also found in garbage dumps, sewers, caves, and tree holes.

Behavior: Breeds mostly after heavy rains from March through August. Males call from flooded meadows and fields, rivers and streams, pools among dunes, flooded ditches, and in a variety of other still-water habitats. Coastal Plain Toads hide by day under logs and other objects and take refuge in rodent burrows. When alarmed, they flatten themselves against the ground.

Voice: The advertisement call is a rattling trill lasting about four to six seconds. It is much less musical and generally lower in pitch than the trill of the American Toad, yet it is distinctly higher in pitch than the rattling trill of the Cane Toad. The male's vocal sac is huge, extending from the lips to the abdomen.

Texas Toad

Bufo speciosus (Anaxyrus speciosus) (2″–3 5/8″)

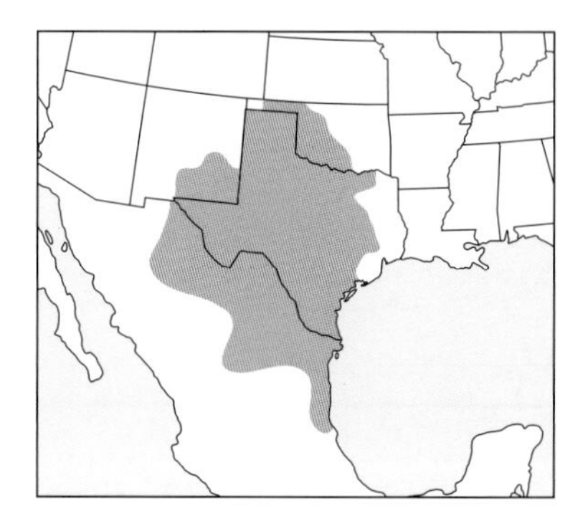

Although the Texas Toad is rather plain in appearance, the raucous high-pitched call of the male is not plain at all and leaves a lasting, and often unfavorable, impression. During dry periods, Texas Toads are rarely seen—they escape heat and drought by burrowing into sand, hiding in gopher burrows, and finding shelter under rocks. They emerge at night to feed, especially after rains.

Appearance: The ground color varies from a bland gray-brown to olive, sometimes with indistinct spots on the back and sides. There is no middorsal stripe, the cranial crests are not well developed, and there are no well-defined dorsal color patterns. As in other true toads, there are two black digging tubercles (spades) on each hind foot, but in the Texas Toad the inner tubercles are sickle or crescent shaped. The male has a large, sausage-shaped vocal sac that extends beyond the nose when inflated and is covered by a flap of skin when deflated.

Range and Habitat: Found throughout much of Texas (except in the easternmost part of the state) as well as western Oklahoma, eastern New Mexico, and adjacent areas in Mexico. Well adapted for dry conditions, especially sandy soils in grasslands, pastures, and open woodlands. Although populations seem to be stable, there is some concern about declines in the Rio Grande Valley of Texas, possibly due to agricultural pesticide and fertilizer runoff.

Behavior: Breeding occurs explosively within several hours of heavy rains, usually from March to September, when males can be heard calling from temporary pools, slow-moving streams, and irrigation ditches.

Voice: The advertisement call of the male is a loud, metallic, rattling trill, lasting about half a second and regularly repeated. It is jarring to the human ear and a large chorus can be overwhelming.

Houston Toad

Bufo houstonensis (Anaxyrus houstonensis) (2″–3 1/8″)

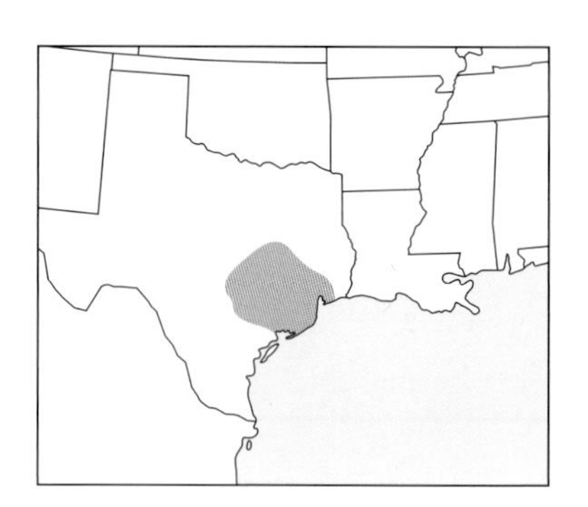

Resembling its close relative the American Toad, the Houston Toad is an endangered species that is actually found in greatest numbers in isolated populations near Bastrop, Texas, rather than near Houston, as its name implies (Houston populations have been extirpated). It looks and sounds much like the American Toad but is considered a Pleistocene relict of that parental species, left behind in suitable habitats in Texas when the American Toad shifted its range north with the melting of the glaciers.

Appearance: Coloration varies from tan or light gray to dark brown, and some individuals may show a tinge of purple, especially on the vocal sacs of males. Houston Toads can be distinguished from American Toads by their range, which is nonoverlapping, and by their somewhat thicker cranial crests (page 126).

Range and Habitat: Houston Toads are found in and around pine forests, especially in Bastrop County (east of Austin); their numbers and reproductive activity have been monitored in Bastrop State Park for many years. They prefer sandy soils in which they can easily burrow. Populations have declined drastically during the last century over much of its range. Efforts to breed them in captivity and release adults, juveniles, and tadpoles have been unsuccessful.

Behavior: Houston Toads are early-spring breeders in permanent and semipermanent ponds, temporary rain pools, flooded fields, and roadside ditches. Most breeding takes place in February and March during warm, humid nights. The breeding season is generally short, lasting only two or three weeks. Each evening, the toads move from their daytime burrows to the breeding site, and then the next day they return to their burrows. Males are usually easy to find as they call from shallow water and protruding grassy patches.

Voice: The advertisement call is a musical trill lasting about fifteen seconds. It sounds very similar to the trill of the American Toad but is slightly higher in pitch.

Cane Toad

Bufo marinus (Rhinella marina) (4″– 7″)

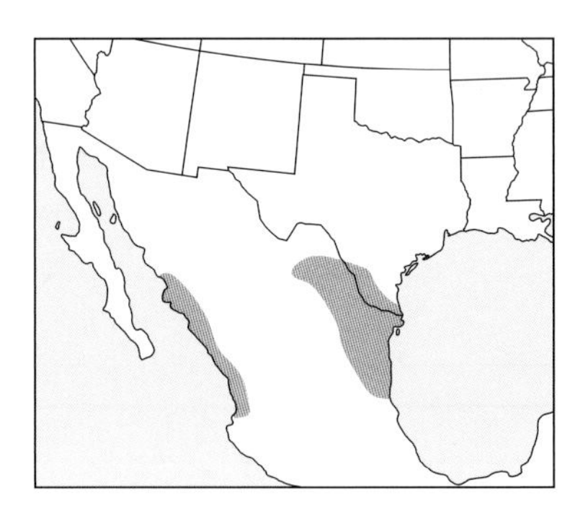

The Cane Toad, also called the Marine or Giant Toad, is one of the largest toads in North America. Adults in the United States grow to nearly seven inches long—only the Sonoran Desert Toad reaches similar lengths. A widespread New World tropical species, Cane Toads have been introduced to a variety of other tropical regions in a futile attempt to control insect pests in sugar-cane plantations. In Australia, their populations have grown to the point that they themselves have become serious pests.

Appearance: Cane Toads are brown to reddish brown in color and have prominent cranial crests. Their parotoid glands are huge, and secretions from these glands are so poisonous that they can kill pets that attempt to eat the toads. In comparison to most other toads of the genus *Bufo,* Cane Toads have smooth skin, but breeding males do develop spine-tipped warts.

Range and Habitat: The natural range extends from the Amazon River Basin of South America north to the lower Rio Grande Valley in Texas. Populations in extreme southern Florida are the result of accidental releases in the Miami area in the late 1950s. Other isolated populations in Florida probably resulted from separate releases. Cane Toads cannot withstand frost, which probably limits northward range expansion. They are generally found near water but may also be seen around houses and in gardens.

Behavior: Cane Toads may breed anytime during the year, but most mating occurs in spring and summer after rains. Breeding sites include freshwater canals, ditches, pools, and ponds, as well as sites with brackish water. Males have relatively small vocal pouches and generally call from the shoreline and shallow water. Adults do most of their feeding during humid nights, and they hide by day in burrows and under rocks and other objects. Cane Toads estivate during dry periods, taking shelter under rocks and logs and burrowing into the soil.

Voice: The advertisement call is a low-pitched rattling trill lasting about four to six seconds. The call of a large individual is noticeably deeper and more resonant than that of a smaller individual, and it may sound like an old one-cylinder engine puttering in the distance.

Great Plains Toad

Bufo cognatus (Anaxyrus cognatus) (1 7/8"–4 1/2")

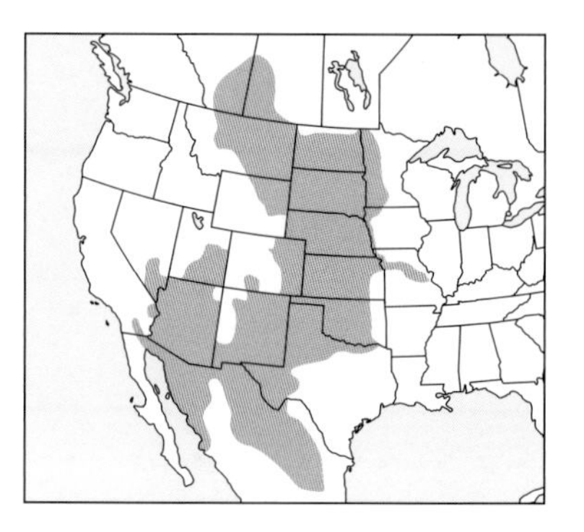

A robust toad with a bold back pattern, the Great Plains Toad is a denizen of open, relatively dry country. Great Plains Toads have spades on their hind feet that make them well equipped for burrowing, and they seldom come to the surface during the day. Their shrill calls are among the longest in duration of any North American frogs or toads, sometimes lasting for almost a minute.

Appearance: The large blotches on the back, each containing many warts, range from light green to gray in color and are set off by surrounding lighter skin. The cranial crests on the back of the head are well separated but converge between the eyes to form a bump, or boss. The male's vocal pouch is large and has a curved, sausagelike shape when inflated.

Range and Habitat: Primarily a prairie grassland species, Great Plains Toads range from Mexico northward through the Great Plains to extreme southern Manitoba and Saskatchewan. They also frequent agricultural land and desert regions of the Southwest.

Behavior: Breeds in river floodplains, ponds, ditches, and flooded fields from April to September, depending on weather and location. Over most of its range, explosive breeding occurs after the first warm spring rains. In desert regions, however, breeding is restricted to seasonal summer rainy periods (monsoons). Satellite behavior is common, and calling males may be surrounded by as many as five, usually smaller, silent males. These males try, sometimes successfully, to intercept females that are attracted by the calling male. Satellite behavior can lead to hybridization in mixed-species choruses because males cannot distinguish between females of their own and other species. A hybrid between a Great Plains Toad and a Woodhouse's Toad has been documented.

Voice: The loud advertisement call is an extended, pulsating trill with a shrill, rattling, metallic quality: *chiga-chiga-chiga-chiga-chiga-chiga.* The call is reminiscent of the sound made by a pneumatic drill and has a rapid pulse rate of about fifteen pulses per second. Choruses can be heard from a great distance and are deafening at close range.

Canadian Toad

Bufo hemiophrys (Anaxyrus hemiophrys) (2″–3 1/4″)

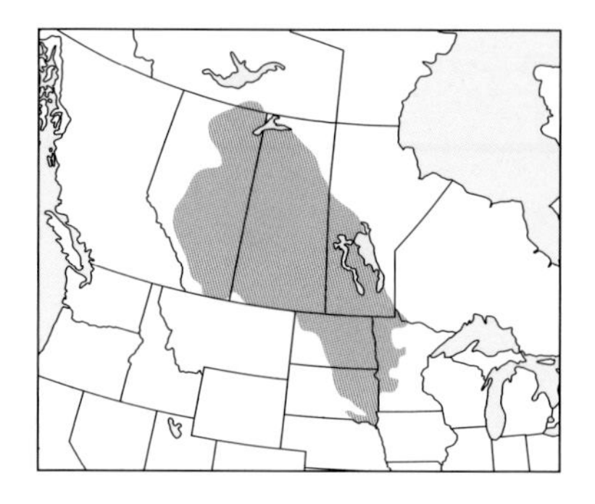

The Canadian Toad is largely confined to Canada but also ranges into the north-central prairie region of the United States. It is more aquatic than most other toads and is often found in and around wetlands. In wet prairie regions of Minnesota, Canadian Toads are known to hibernate in large mounds of loose soil, called Mima mounds, which are probably created by the earthmoving activities of pocket gophers and other vertebrates. Such mounds can contain thousands of hibernating toads.

Appearance: This species is unique among North American toads in that the adult has a raised area or boss on its head that begins on the snout and extends along the midline to the back of the eyes. The boss is often grooved on top and probably results from a fusion of the cranial crests (page 126). The dorsal ground color ranges from olive to reddish, and there are many small brown or reddish warts that are concentrated in dark spots. There is a light stripe running along the center of the back.

Range and Habitat: Common in the central Canadian provinces, including the Northwest Territories, Alberta, Saskatchewan, and Manitoba. In the United States, ranges over most of North Dakota and also into northwestern Minnesota and the eastern third of South Dakota. Inhabits the borders of lakes, potholes, and other bodies of permanent water in prairies and open aspen groves.

Behavior: The Canadian Toad breeds from early spring to late summer, depending on the latitude, altitude, and local climate. In some localities it is more diurnal during the breeding season than other toads, taking refuge at night in self-made burrows. When frightened, it is more likely than most toads to escape by swimming into deep water. Frequents prairies and open parklands outside the breeding season but generally stays close to water.

Voice: The advertisement call is a fairly melodic trill lasting about three to six seconds. It is softer and not quite as musical as the trill of the American Toad, with which it hybridizes at the southeastern edge of its range.

Wyoming Toad

Bufo baxteri (Anaxyrus baxteri) (1⅞″–2⅜″)

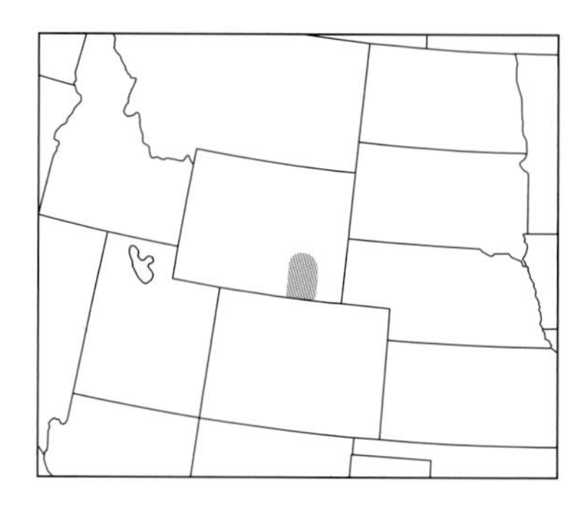

Long considered to be a subspecies of the Canadian Toad, the Wyoming Toad represents a group of high-elevation relict populations of Canadian Toads that became isolated because of warming at the end of the Pleistocene glaciation. They were abundant in river floodplains near Laramie in the 1950s but became extinct in the wild by the mid-1990s, probably because of infections by chytrid fungus (page 35). A single reintroduced population currently exists, founded from captive-breeding programs at zoos.

Appearance: The ground color may be dark brown, gray, or greenish with dark blotches and an indistinct light line down the middle of the back. Warts are large and prominent. Like the Canadian Toad, the Wyoming Toad has cranial crests that are fused along the midline to form an elongate boss that is grooved on top. Wyoming Toads are smaller than Canadian Toads, with an average length of around two inches.

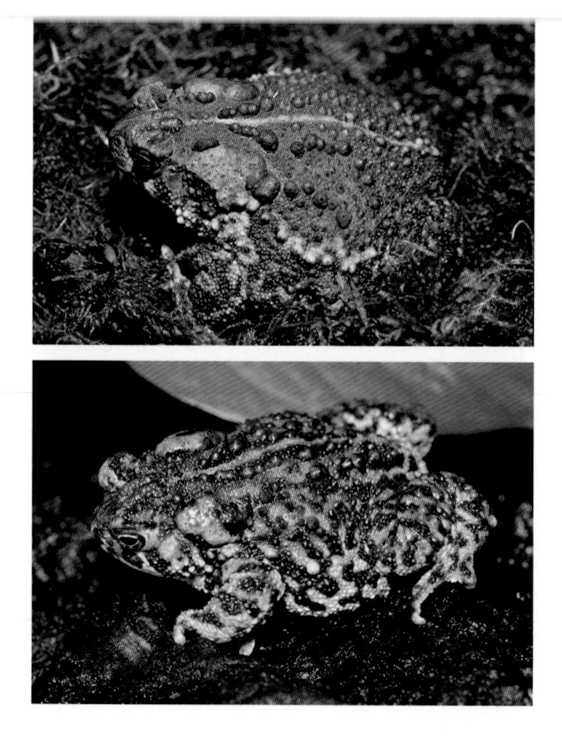

Range and Habitat: The Wyoming Toad is currently restricted to Mortenson Lake, Wyoming, where a single introduced population is barely holding its own against the chytrid fungus that poses an ongoing threat. Captive breeding continues in an effort to save the species. Like the Canadian Toad, the Wyoming Toad is attracted to water and inhabits the borders of lakes and other bodies of permanent water in prairies and open aspen groves.

Behavior: The habits of Wyoming Toads are similar to those of Canadian Toads. In wild populations, they breed from spring to early summer. They frequent short-grass prairie along the edges of lakes, ponds, streams, and floodplains and are rarely found more than thirty feet from water. When alarmed, they dive into open water and swim to the bottom.

Voice: The advertisement call is a soft, fairly melodic trill lasting from three to five seconds. It has a rattling quality and is not as musical as the melodic trill of the American Toad.

Red-spotted Toad

Bufo punctatus (Anaxyrus punctatus) (1½″–3″)

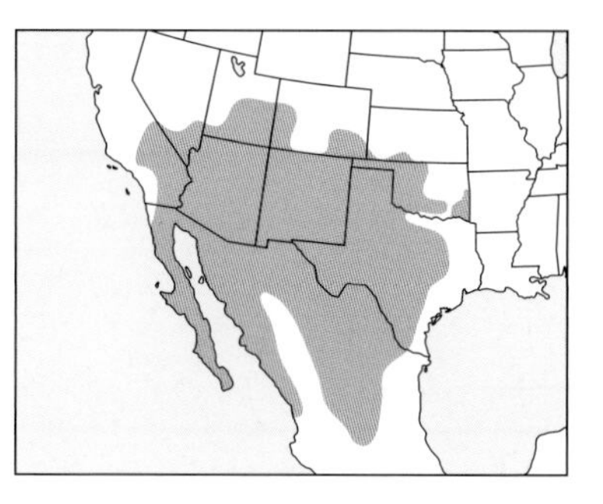

The Red-spotted Toad of the arid Southwest is named for the red or orange spots that are usually abundantly scattered on its back, sides, and legs. Common in rocky areas, they are good climbers and their small, flat bodies allow them to easily hide under rocks and in crevices and cracks.

Appearance: The ground color may be gray, tan, olive, or brown. With or without spots, the species can be recognized by its distinctly round parotoid glands (they are generally elongate in other toads), which are about the same size as their eyes. The cranial crests are either absent or poorly defined (page 126). Spots may be absent in large adults.

Range and Habitat: Ranges from the western half of Texas to southeastern California and north into Utah, Colorado, southwestern Kansas, and Oklahoma (also widespread in northern Mexico). Found in a variety of arid and semiarid habitats from coniferous forests and high grasslands to lowland desert, where it is locally abundant near rocky streams, pools, springs, cattle tanks, and other water sources.

Behavior: Breeding takes place from March to September, independent of rains in the spring and during and after heavy rains in the summer. Individuals migrate to small rocky creeks, pools, and springs—wherever ample water exists. Males typically call from shallow water along the shoreline and from the tops of rocks at the water's edge. Occasionally, they call from burrows, from rock crevices, and from under rocks near the water's edge.

Voice: The advertisement call is a high-pitched musical trill that lasts about three to ten seconds. It resembles the call of the Arizona Toad but is higher in pitch. The pulse rate of trills is rapid, and distant calls might easily be mistaken for the soft trills of tree crickets.

Green Toad

Bufo debilis (Anaxyrus debilis) (1 1/8"–2")

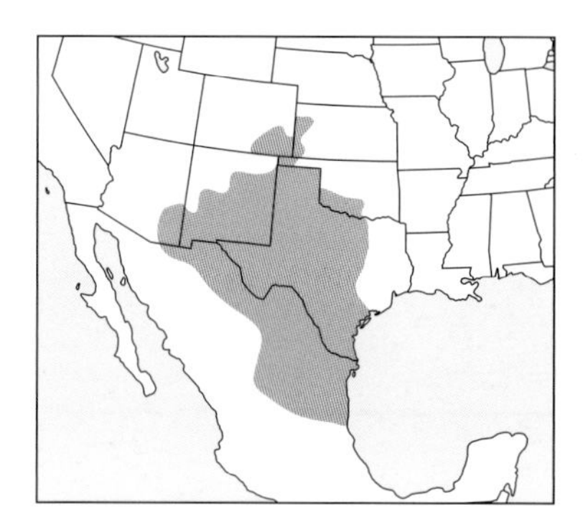

Smallish and flat-bodied with a dorsal coloring of green, yellow, and black, the secretive Green Toad is one of our most attractive toads. Although abundant in many parts of their range, Green Toads are nocturnal and seldom seen in their prairie and desert habitats, where adults take refuge in rodent burrows and hide camouflaged among the grasses.

Appearance: There are two subspecies. The Eastern Green Toad *(B. debilis debilis)* usually has a very bright green ground color and many small and unconnected black and yellow spots. In contrast, the Western Green Toad *(B. debilis insidior)* is generally pale green or greenish yellow with black lines and spots that are sometimes connected, although not in a clear netlike pattern as in the Sonoran Green Toad (*B. retiformis;* page 166). In both subspecies, the ventral sides are cream or white, and breeding males have dusky throats. The parotoid glands are elliptical, and cranial crests are low or absent.

Range and Habitat: Found from Texas north into southwestern Oklahoma and extreme western Kansas, and then westward into Colorado, New Mexico, and southeastern Arizona (as well as in northeastern Mexico). The line separating the two subspecies runs longitudinally through western Texas. Green Toads favor grassland habitats from prairie to desert but are also found in mesquite and creosote-bush scrub. The Western subspecies is especially well adapted to arid and semiarid habitats.

Behavior: Breeding takes place after heavy rains from March through August in temporary pools and ditches, as well as in cattle tanks and slow-moving sections of creeks.

Voice: The advertisement call is a penetrating buzzy trill lasting from three to ten seconds, with pauses of several seconds or longer between trills; sometimes described as sounding like an electric buzzer or a cricket. The nasal call sounds similar to that of the Great Plains Narrowmouth Toad, a species with which the Green Toad widely overlaps in range (page 280).

Sonoran Green Toad

Bufo retiformis (Anaxyrus retiformis) (1 1/8"–2 4/5")

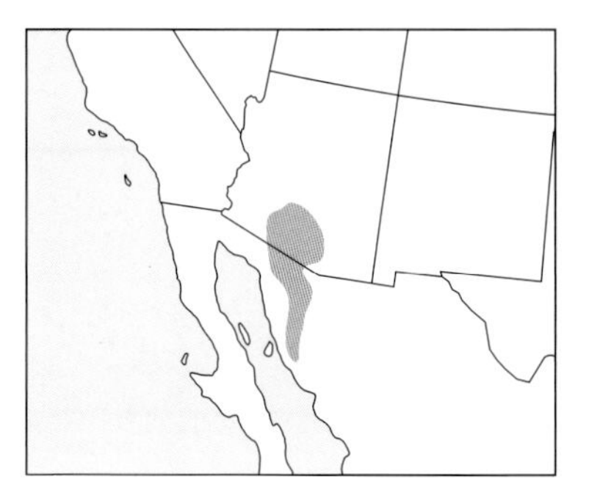

A small toad native to semiarid habitats in south-central Arizona, the brilliantly colored Sonoran Green Toad is nocturnal and remains hidden underground during most of the year. While it may come out to feed on humid nights, it is best observed when large numbers emerge to breed after summer monsoon rains.

Appearance: Its coloration is striking—there is a netlike pattern of black or dark brown on the back that frames a profusion of brilliant yellow, yellow-green, or green oval spots. The parotoid glands are large, elongate, and covered with tiny black-tipped warts. The cranial crests are absent or reduced.

Range and Habitat: Found in Sonoran Desert scrub, semidesert grasslands, and thorn scrub from south-central Arizona to central Sonora, Mexico. Frequents valleys, arroyo bottomlands, playas, roadside ditches, and cattle tanks that fill during the summer rainy season. Sonoran Green Toads are often associated with mesquite thickets. The range may be expanding northward because of agricultural activities and other disturbances that create places where water accumulates.

Behavior: The Sonoran Green Toad usually breeds explosively within a day or two of the arrival of summer monsoon rains in July or August. Males gather and call in grassy and shrubby areas near temporary pools. Males mount approaching females, which then carry the males to the breeding pools where the eggs are fertilized and laid.

Voice: The call is a dissonant nasal buzz (with an uneven beginning) that lasts about three seconds and is repeated every few seconds or so: *waaaaaaaaaa.* It sounds similar to the nasal calls of the Great Plains Narrowmouth Toad, which occurs within its range, but is typically hoarser and lower in pitch. The buzz can be deceptive, often seeming farther away than it really is.

Sonoran Desert Toad

Bufo alvarius (Ollotis alvaria) (4"–7½")

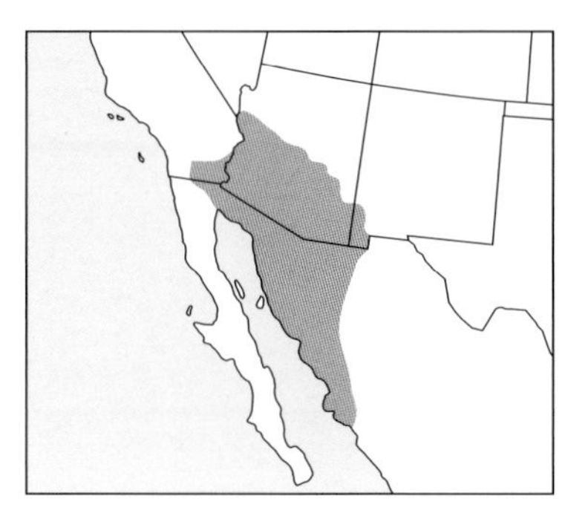

Capable of reaching lengths of over seven inches, the Sonoran Desert Toad rivals the Cane Toad as the largest of our native toads. Also known as the Colorado River Toad, it is a southwestern species common throughout much of lowland Arizona. Venom from the parotoid glands is strong enough to paralyze or kill a dog. The venom also possesses hallucinogenic properties, and some people get high by licking the skin around the glands—a dangerous practice referred to as toad licking—or by smoking the dried venom or skin.

Appearance: Resembling a lump of cow dung when observed from a distance, the adult Sonoran Desert Toad is relatively smooth skinned and uniformly colored from olive to brown or gray. Cranial crests are prominent, and the parotoid glands are elongate. There is a round whitish wart or turbercle at the angle of the jaw, prominent large glandular swellings on the hind legs, and small orange-colored tubercles scattered on the back and sides (tubercles are particularly prominent on light colored juveniles).

Adult (top) and juvenile (bottom)

Range and Habitat: Found in the United States from extreme southwestern New Mexico through the southern half of Arizona and to extreme southeastern California (where it may now be extirpated). The species also occurs throughout much of the state of Sonora, in Mexico. Desert and mesquite scrub are its favored habitats, but it may also be found in grasslands, riparian zones along rivers, and occasionally in lower wooded canyons and slopes.

Behavior: Sonoran Desert Toads feed at night and hide by day in rodent burrows and under rocks and other objects. They breed from May to July in pools, ditches, springs, cattle tanks, ponds, canals, and reservoirs, usually following heavy summer monsoon rains. Males may call by themselves or in small choruses. Breeding usually lasts only a night or two but may be extended in areas where permanent water exists. When not breeding, they wander far from water and can move surprisingly fast.

Voice: The advertisement call is a quiet, garbled *waahhh,* lasting less than a second and given once every few seconds. It sounds similar to the nasal cry of Woodhouse's Toad but is shorter and softer.

Arizona Toad

Bufo microscaphus (Anaxyrus microscaphus) (2″–3⅛″)

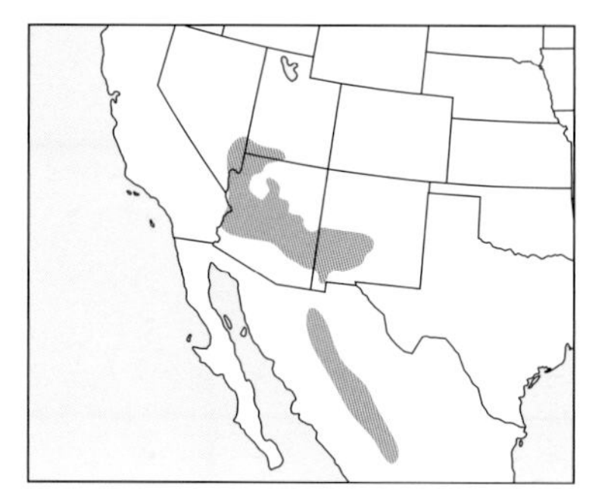

The Arizona Toad is common across central Arizona but also ranges into the surrounding states of Nevada, Utah, and New Mexico. In areas of overlap with Woodhouse's Toads, the two species often hybridize, especially where damming, irrigation, and other human disturbances have created still-water habitats favored by Woodhouse's Toads. At such locations, Arizona Toads tend to lose their distinctiveness as they interbreed with the more abundant Woodhouse's Toads.

Appearance: The dorsal color is often gray but may vary from brown or rust to pink or yellowish. As in its close relative the Arroyo Toad (the two were long considered to be subspecies), the parotoid glands are small and oval, cranial crests are absent, a dorsal stripe is weak or absent, a light band usually exists between the eyes, and there is sometimes a light-colored area (or two) on the back.

Range and Habitat: Found in riparian areas from the Colorado and Virgin River basins in extreme southwestern Utah and southern Nevada through the lower mountains of central Arizona to the Mogollon Rim in Arizona and southwestern New Mexico. Frequents shallow, sandy seasonal and permanent streams in arid lowlands, as well as rocky streams in mountainous areas with oak or pine forests.

Behavior: Arizona Toads breed from February to June or July, depending on altitude and local weather conditions. Rainfall is not necessary to initiate breeding, which may be brief, lasting only a week or two. Males often call from rocks and sandbars along the slow-moving parts of streams and in pools near streams.

Voice: The advertisement call is a long musical trill lasting about six to ten seconds. Trills usually start soft and rise slightly in pitch at the beginning, before becoming loud and steady.

Western Toad

Bufo boreas (Anaxyrus boreas) (2″– 5″)

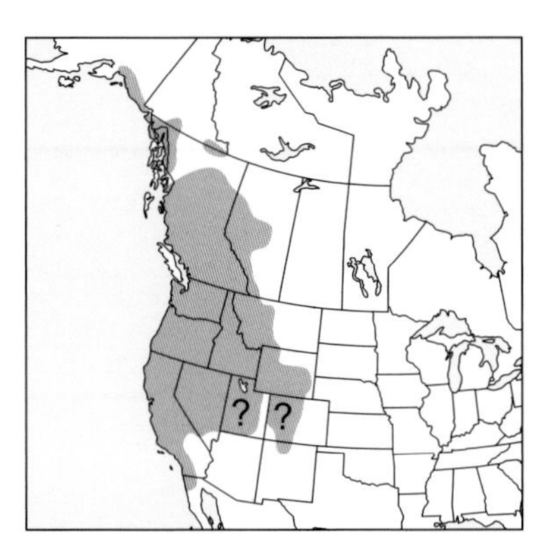

More likely to walk than hop, the wide-ranging and variously colored Western Toad has experienced major population declines in certain parts of its range (see below). There are two recognized subspecies, the Boreal Toad *(B. boreas boreas)* and the California Toad *(B. boreas halophilus).* Many believe that the complex of diverse populations we now label as Western Toad will ultimately be divided into several different species.

Appearance: The Western Toad lacks cranial crests and usually has a pale white or yellow dorsal stripe. The ground color may be blackish, gray, brown, reddish, greenish, yellowish, or tan. The back and sides are covered with brown or reddish warts or tubercles, each surrounded by a black border. Females typically have smoother skin and more complex color patterns than males. The two subspecies differ somewhat in appearance. Boreal Toads are relatively dark above and usually have considerable blotching on the undersides. In contrast, California Toads are usually lighter above and have less dark blotching below. Western Toads generally lack vocal sacs.

Range and Habitat: The Boreal Toad occurs from southeastern Alaska to northern California and inland to western Colorado, Wyoming, and Montana. The California Toad ranges from northern California south into Baja California. Preferred habitats of both subspecies are diverse, ranging from mountain meadows and woodlands to desert springs and streams. Severe population declines have occurred throughout the southern Rocky Mountains in Colorado and Wyoming. Suspected causes include habitat degradation and chytrid fungus infections.

Behavior: The time of breeding depends on latitude, elevation, and local weather conditions. It may occur as early as January or February in southern portions of the range and as late as August at high elevations in the Rocky Mountains of Colorado. Males call from a variety of temporary, semipermanent, and permanent bodies of water, all characterized by having little or no flow.

Voice: The call of the Western Toad is a birdlike chirping or peeping sound. Since males call primarily when surrounded by other males, some believe the calls function as release calls or short-range aggressive calls rather than as advertisement calls.

— Note: Unless otherwise indicated, photos depict color variations of the Boreal Toad subspecies (Bufo boreas boreas) *—*

California Toad subspecies
(Bufo boreas halophilus)

Amargosa Toad

Bufo nelsoni (Anaxyrus nelsoni) (2″– 4 2/5″)

Previously considered a subspecies of the Western Toad, the Amargosa Toad is restricted in range to a small area in south-central Nevada on the edge of the Amargosa Desert. The scenario that led to its isolation and differentiation is a classic example of evolution. As the Great Basin gradually dried during the late Pleistocene, populations of the Western Toad became separated by unsuitable arid habitat but continued to survive and reproduce in isolated desert drainages. Two of these so-called desert isolates, the Amargosa Toad and the Black Toad, eventually became species, changing and adapting to their particular environments over many thousands of years.

Appearance: The Amargosa Toad closely resembles the Western Toad. But whereas the Western Toad may be highly variable in color, the Amargosa Toad consistently has a light brown or tan ground color with orangish warts surrounded by bold black borders. As in the Western Toad, the parotoid glands are oval, and a prominent light middorsal stripe is present.

Range and Habitat: The Amargosa Toad is found only in the Oasis Valley of Nevada, near the town of Beatty, along a ten-mile stretch of the Amargosa River and associated smaller streams and springs. The population seems to be stable at around 20,000 individuals, but its tiny range makes it particularly susceptible to effects of habitat modification, water diversion, disease, and the like. It is currently listed as a species of special concern, but there is no federal protection of either the species or its habitat. The State of Nevada lists the Amargosa Toad as a protected species, and citizens of the town of Beatty, along with a number of local residents, are active in protecting and restoring habitats where the toad is found.

Behavior: May be found breeding from February into summer in slow-moving water, shallows, and springs. Takes refuge by day in burrows, under debris, and in dense vegetation.

Voice: The Amargosa Toad lacks a vocal sac and is mostly mute, but it does give chirplike release calls when grasped by another toad or a human. These release calls closely resemble the chirps of Western Toads.

Black Toad

Bufo exsul (Anaxyrus exsul) (1 3/4″–3″)

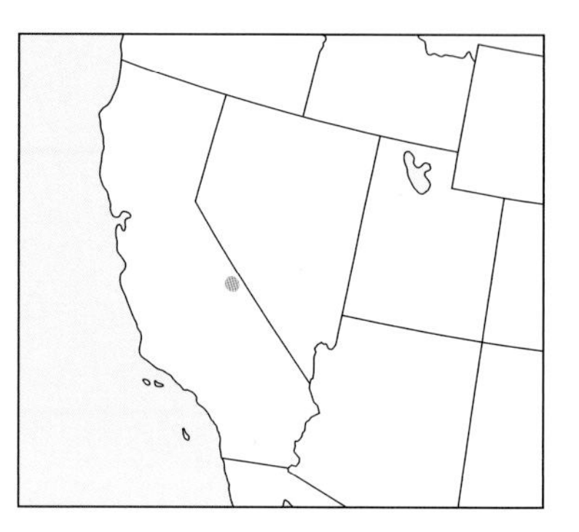

Occupying a tiny range along the east-central border of California, the Black Toad has evolved as a desert isolate, completely surrounded by dry country that prevented, and still prevents, interbreeding with its close relative the Western Toad. Its specific name, *exsul,* is Latin for "in exile," which refers to its geographic isolation, a result of widespread warming at the end of the Pleistocene.

Appearance: The Black Toad gets its common name from its black dorsal color, which is adorned with scattered white flecks or jagged lines and a prominent white middorsal stripe. The ventral surface is light with dark spots or blotches. Black Toads are slightly smaller in size than Western Toads.

Range and Habitat: Black Toads are confined to about ten different spring systems in Deep Springs Valley, which is located in an isolated desert basin in east-central California adjacent to the Nevada border. (There is also a small introduced population at a spring-fed well in Death Valley National Park.) Deep Springs College, which owns most of the land on which the toads are found, is actively involved in conservation efforts. The total number of individuals is estimated to be around 20,000, and populations appear to be stable.

Behavior: Black Toads are largely aquatic, prefer wet areas with lots of vegetation, and rarely wander more than about thirty feet from the water. Breeding occurs from mid-March to May in marshes, ponds, and shallow pools associated with spring systems. Adults are primarily diurnal in habits but may become active at night during hot weather.

Voice: Males lack vocal sacs, but when the toads are clasped or harassed by other males they give repeated chirplike calls that sound basically the same as the calls made by Western Toads. These calls probably function as release or aggressive calls rather than as advertisement calls.

Arroyo Toad

Bufo californicus (Anaxyrus californicus) (1 4/5″–3 2/5″)

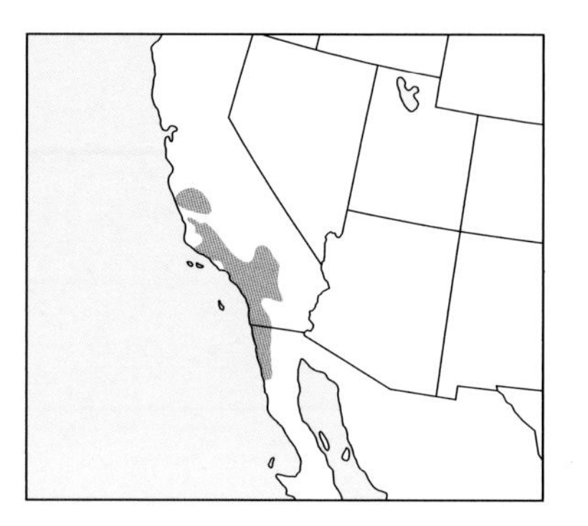

Formerly considered a subspecies of the Arizona Toad, Arroyo Toads are confined to elevations below about 3,000 feet in southern California and the Baja peninsula. They are named after one of their most frequented habitats, the arroyo, a steep-sided watercourse with a nearly flat floor that is often dry except after heavy rains. The Arroyo Toad is an endangered species, and many populations have been extirpated in recent years.

Appearance: Arroyo Toads are stocky and uniformly covered with warts that have brown-tipped projections. Cranial crests are poorly developed or absent. The parotoid glands are widely separated, oval in shape, and pale toward the front. The ground color is typically gray, olive, or brown, and there is usually a light-colored band between the eyes and a light area on the middle of the back.

Range and Habitat: Mainly occurs in coastal mountains and valleys of California, from southern Monterey County to northwestern Baja California. Some populations are associated with larger streams on the desert slopes of the San Gabriel and San Bernardino mountain ranges. Arroyo Toads live in and around sandy-bottomed seasonal and permanent streams that are subject to flash floods and that often have braided channels. Primarily because of human water-management practices coupled with several decades of severe drought, Arroyo Toads were extirpated from many of their historic locations from the 1950s through the 1990s. They were listed as endangered in 1994. Improved water-management practices over the past decade have resulted in some recovery, especially on public lands.

Behavior: Breeding is prolonged, beginning in February to early April, depending on elevation, and extending well into the summer. Males call from relatively quiet stretches of open streams. Eggs are in jeopardy from both drying of streams and flash floods. In some localities, arid and rocky surroundings restrict terrestrial movements, but in other places, Arroyo Toads visit grasslands and sage scrub up to a mile away from their breeding sites.

Voice: The advertisement call is a long musical trill lasting about six to ten seconds. Trills usually start softly and rise slightly in pitch at the beginning, then become loud and steady. The release call is a rapid series of chirping notes.

52

Yosemite Toad

Bufo canorus (Anaxyrus canorus) (1 3/4"–2 3/4")

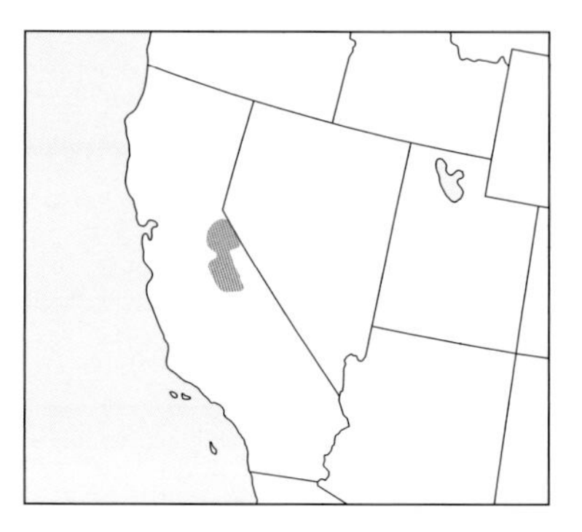

Yosemite Toads inhabit high-elevation mountain meadows and wetlands in the Sierra Nevada Mountains of California, including those in Yosemite National Park. Breeding males produce pleasant trills that can be heard from hundreds of feet away—a musical utterance that earned the species its name, *canorus,* which is Latin for "tuneful."

Appearance: Compared to Western Toads, Yosemite Toads have smoother skin, proportionally larger parotoid glands, and more closely set eyes. The sexes differ greatly in appearance. Females are ornamented with a profusion of dark spots or blotches, often edged with white, set against a light brown or tan ground color. By comparison, the males are drab and more evenly colored. Their ground color ranges from dark olive to pale brown or yellow-green, sometimes with a scattering of small dark spots or flecks.

Range and Habitat: Yosemite Toads occur in the high Sierra Nevadas of California from the vicinity of Ebbets Pass (about twenty-five miles north of Yosemite National Park) southward to the Spanish Mountain area in Kings Canyon National Park. Found mostly at elevations of about 8,000 to 10,000 feet, where it frequents mountain meadows and forest edges. The species hybridizes with Western Toads in places where their ranges overlap. Recent studies have documented sharp declines in many populations; disease may be the cause in some of these cases.

Behavior: Yosemite Toads are diurnal in habits. Breeding occurs from May through August, usually in shallow pools filled by snowmelt and in small streams in and around meadows. Males have large vocal sacs and call during the day. Often, many more males than females are present at breeding sites (ten times more males, in one study). Outside the breeding season, adults often move into densely vegetated areas at the edges of meadows to feed until the arrival of snow. Yosemite Toads are known to take shelter and overwinter in the abandoned burrows of a variety of mammals.

Ornately colored female

Voice: The advertisement call is a pleasing, melodic trill, sometimes with a slight rattling quality. Trills last about five seconds and are repeated every few seconds.

Drably colored males

True Frogs — Family Ranidae

Members of the family Ranidae are referred to as true frogs or, more simply, ranids. They are often described as typical-looking frogs that are medium to large in size, with long legs and webbing between the toes. Most species frequent moist areas next to lakes, ponds, and streams, and startled individuals often leap long distances to safety in the water. Nearly worldwide in distribution, the family includes 650 species and 38 genera. In North America, 28 species are currently classified as members of the genus *Rana* (but see last paragraph). The group includes some of our most familiar and well-known frogs, such as the Bullfrog, Green Frog, and various leopard frogs. One extinct species, the Vegas Valley Leopard Frog *(Rana fisheri),* once inhabited springs near Las Vegas but has not been seen since 1942.

Pickerel Frog

Most of our true frogs live in or near water, with the exception of the Gopher Frog and Crawfish Frog, which spend much of their lives in burrows. While most species are medium-sized, there is considerable variation in length within the group. The wide-ranging Bullfrog is our largest ranid, reaching lengths of nearly eight inches. At the other end of the spectrum is the diminutive Florida Bog Frog, less than two inches long and confined to one small area of the Florida Panhandle. True frogs usually breed in semipermanent to permanent bodies of water, and their tadpoles may take years to develop. Several species, including the Bullfrog and Green Frog, are territorial, with males defending areas that contain good habitat for females to lay eggs.

Aside from a handful of western species, most of our North American species do not show close affinities to Eurasian members of the genus *Rana.* For this reason, it was recently proposed that these be placed in a different genus, *Lithobates.* If that proposal is accepted by the community of taxonomists, scientific names for the majority of ranids in the United States and Canada will change. The downside to this change—and to a similar generic change within the family Bufonidae (page 127)—is that it will cause confusion in the vast scientific literature about these frogs, and so many authors will continue to use the old generic designations in order to provide continuity. In anticipation of this change, we include the alternative names in parentheses in the headings of the species accounts. In all likelihood, the common names will remain unchanged.

Green Frog

Bullfrog

Rana catesbeiana (Lithobates catesbeianus) (3½″–8″)

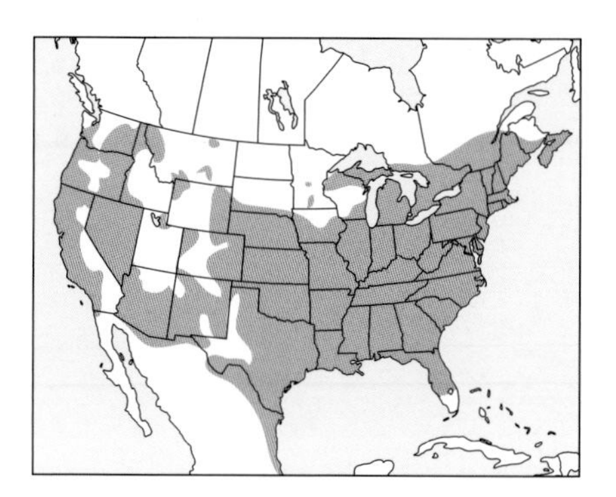

Our largest and most familiar native frog, the Bullfrog is common throughout the eastern United States. Introduced in much of western North America (and other parts of the world), it poses a serious threat to native frogs because it will eat almost anything smaller than itself. Bullfrogs are named for their advertisement call, which is likened to the bellowing of a bull.

Appearance: Bullfrogs are variable in color, but usually are entirely green above and always have green heads. Especially in the Southeast, individuals may have reticulated patterns of gray, brown, or rust (see photo on page 189). The underside is whitish with a yellow tinge, often with dark markings that are less prominent than in the Pig Frog. Bullfrogs also differ from Pig Frogs in having less extensive hind-foot webbing and a blunter snout. Dorsolateral folds are lacking, a characteristic that serves to distinguish Bullfrogs from Green and Bronze Frogs. Breeding males have yellow throats (see photo on page 188).

Range and Habitat: Found throughout the eastern states (except southern Florida and the upper Midwest) and west through most of Texas, Oklahoma, Kansas, and Nebraska. Introduced populations occur in nearly all the western states and in southern British Columbia. Prefers large permanent bodies of water such as lakes, ponds, and slow-moving streams, where it is usually found in the water or along the shoreline. The Bullfrog is declining in some parts of its native range because of overharvesting for frog legs, pollution, introduced diseases, and habitat destruction.

Behavior: In spring and early summer, males establish territories by calling and having wrestling bouts. Territories soon stabilize, and males learn to recognize the calls of nearby males, with neighbors calling back and forth in an alternating fashion. But if the call of an unfamiliar male is heard, aggressive behavior immediately resumes.

Because only large males are likely to obtain territories, smaller males either call opportunistically and retreat when challenged or adopt the satellite tactic, remaining silent and trying to intercept incoming females (page 24).

Voice: The advertisement call is a series of loud, resonant bass notes sounding like *rumm . . . rumm . . . rumm* or a stuttering *ru-u-u-ummm . . . ru-u-u-ummm* (often transliterated as "jug-o-rum"). The aggressive call is an abrupt, spitlike *phphoot!* A frightened individual, especially a juvenile, may give a loud *eeek!* when it leaps into the water.

Green Frog

Rana clamitans (Lithobates clamitans) (2 1/4"–4 1/4")

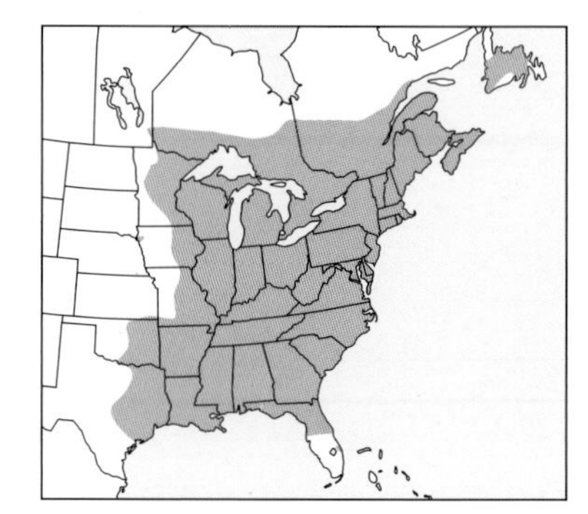

Enlivening ponds, lakes, and swamps with banjolike *gunk!* calls, Green Frogs and Bronze Frogs are often mistaken for small Bullfrogs as they squeak and leap into the water when flushed from the shoreline. Over much of eastern North America, when a child goes to a pond to find a frog, this species is the one most likely to be encountered.

Appearance: The northern subspecies, referred to as the Green Frog *(R. clamitans melanota),* is often more brown than green and has a scattering of dark brown or blackish spots (blue individuals are sometimes found, their color resulting from a lack of yellow pigment). The southeastern subspecies, the Bronze Frog *(R. clamitans clamitans),* is named for its bronzy dorsal color (pictured on pages 192–193). Individuals of both subspecies typically have light green upper lips and well-developed dorsolateral folds, making them easy to distinguish from Bullfrogs, River Frogs, and Pig Frogs, which lack such folds. Adult males have very large eardrums, distinctly larger than their eyes.

Range and Habitat: Green Frogs are found mainly in the eastern third of the United States and southeastern Canada but range as far west as eastern Texas and Oklahoma. Bronze Frogs occupy the southeastern and Gulf coastal plain regions and northern Florida and also the Mississippi River drainage north to southern Missouri and Illinois. While Green Frogs frequent almost any shallow, permanent freshwater habitat, Bronze Frogs are most common in swamps and small creeks.

Behavior: Green and Bronze Frogs breed from March to August, depending on the latitude and local climate. As in Bullfrogs, males establish territories in which females lay eggs, and territorial defense may involve wrestling bouts. Males use the pitch of the calls of their rivals to assess their opponents' size during encounters, and a small male may retreat in the presence of a large male, which has a lower-pitched call.

A melanistic Green Frog lacking yellow pigment

Voice: The advertisement (and territorial) call is an explosive, throaty *gunk!* that resembles the sound made by plucking a loose banjo string. Calls are usually delivered in a short series, dropping slightly in pitch and volume from beginning to end: *GUNK!-Gunk!-gunk!* During an encounter, a male may give a stuttering series of guttural notes that have a Bullfrog-like quality, followed by a sharp staccato note. An abrupt *iCUP!* may also be given, and frightened individuals often squeak or chirp when leaping into water.

Color variants of Bronze Frogs (female above, male to the right)

Pig Frog

Rana grylio (Lithobates grylio) (3 1/4″–6 3/8″)

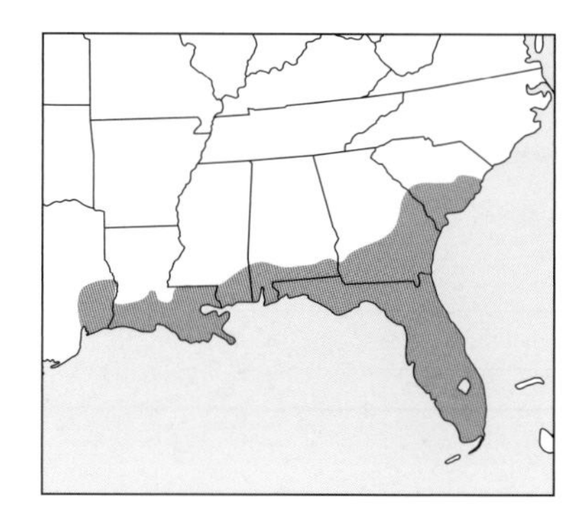

Often confused with the Bullfrog, the Pig Frog is a true southerner, inhabiting lowland swamps and lakes from South Carolina to Texas. It is named for its hoglike call, which is a low-pitched grunt or series of grunts. Huge choruses are impressive to hear. Pig Frogs can grow quite large, with individuals occasionally reaching lengths of six inches or more.

Appearance: The Pig Frog has a narrow and pointed head and more extensive webbing on the hind feet, especially between the third and fourth toes, than any other ranid. Ground color is olive to dark brown, usually with numerous dark spots, and there are no dorsolateral folds. The underside is white or pale yellow, with darker netlike markings on the thighs and lower abdomen. Usually, the backs of the thighs have distinct dark and light bands, although some individuals are more uniformly colored.

Range and Habitat: Restricted to the southern coastal plain and peninsular Florida, Pig Frogs range from Louisiana to South Carolina, where they occur in swamps, marshes, lakes, and large ponds. Much less likely to venture on land than other ranids.

Behavior: Pig Frogs may be heard calling in every month of the year in the southern part of their range, but most breeding is concentrated in the period from March to September. Choruses can be very dense and loud. Tadpoles are large and in some populations require two years to metamorphose. Little is known about their breeding behavior. Males are likely to be territorial, as Bullfrogs are, but this has not been confirmed by observations or experiments. It would be interesting to learn if and how Pig Frogs and Bullfrogs compete. Do they avoid contact by virtue of their different habitat requirements? Do they show interspecific territoriality?

Voice: The Pig Frog's advertisement call is a low-pitched, guttural, piglike grunt, usually repeated two or three times and occasionally (when the temperatures are high) up to seven or more times. A spitlike *phooot!*, similar to that of the Bullfrog, is given during aggressive encounters. Males often call from offshore vegetation, sometimes while floating in the water.

Mink Frog

Rana septentrionalis (Lithobates septentrionalis) (1 7/8"– 3")

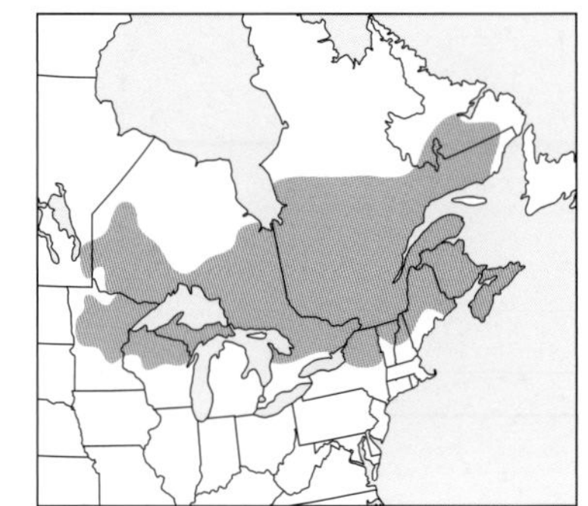

The Mink Frog gets its common name from its musky odor when handled, although the smell has also been likened to that of rotten onions. An inhabitant of northern lakes and bogs, it reveals its presence with a unique advertisement call that resembles the sound made by striking two sticks together.

Appearance: The Mink Frog's dorsal ground color ranges from dark green to brown or olive, and there are usually many spots, blotches, or a netlike pattern of dark markings. The upper jaw and sometimes the forward part of the back is green. Blotches on the upper sides of the hind legs are usually roughly rectangular. The hind feet have webs that extend to the large joints of the fourth toes. Undersides are whitish except for the throat, which ranges from pale to bright yellow (males have the brightest throats). Dorsolateral folds are absent in many individuals but may be partially developed or well developed in others.

Range and Habitat: The bulk of the range is in Canada from Manitoba to Labrador and Newfoundland, but it also extends southward into the United States in northern Minnesota, Wisconsin, the upper peninsula of Michigan, New York, and upper New England. Mink Frogs are highly aquatic, favoring cold water in permanent northern lakes and ponds where there is heavy vegetation. They also occur in the quiet parts of rivers.

Behavior: Mink Frogs breed from late May to August in the same places where they live year-round. The most active calling occurs on warm nights in midsummer, a couple of hours before dawn, when huge choruses can be heard tapping away in boggy north-woods ponds. Individuals will bask in the sun, but they often change their posture and position to control their body temperature. Mink Frogs are intolerant of warm water, however, and this factor (along with avoiding competition with Bullfrogs) has been hypothesized to account for their northern distribution.

Voice: The advertisement call is a series of about four sharp woody raps, *cut-cut-cut-cut.* Aroused individuals produce rolling, stuttered series of calls: *grrruut-grrruut-grrruut-grrruut.*

Wood Frog

Rana sylvatica (Lithobates sylvaticus) (1 3/8"–3 1/4")

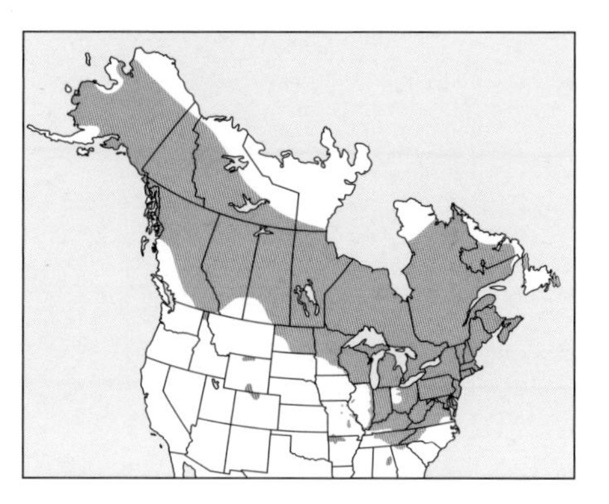

Although showing a marked preference for woodland habitats over most of its range, the Wood Frog is found farther north than any other member of its family, all the way to treeless marshy areas in taiga and tundra settings. They are the only North American amphibian species found north of the Arctic Circle, and they possess cryoprotectant chemicals that protect them from freezing to death. Wood Frogs breed shortly after the snow melts, and their ducklike cackling calls betray their whereabouts in shallow breeding pools.

Appearance: The Wood Frog ranges in color from brown or pinkish brown to copper or tan. There is a prominent dark mask that begins thinly at the nose and then widens as it runs through each eye and extends to include the eardrums. The mask may, however, be hard to see on dark-colored individuals. Dorsolateral folds are prominent, and the groin may have a yellow cast.

Range and Habitat: Ranges from Georgia to Colorado and northward to Labrador, the Yukon, and Alaska. Southern populations are generally restricted to moist woodlands in mountainous or hilly regions. In northern areas, Wood Frogs also frequent woodland habitats but may be found in marshy areas and wet meadows.

Behavior: Wood Frogs are among the first of the spring breeders and may be heard as early as February in southern parts of their range. Indeed, males may start calling as soon as the ice begins to melt. They typically breed in temporary vernal pools that lack predatory fish. The breeding season is brief, usually lasting a few days to a week. Mates are obtained by what has been termed scramble competition. Each male continually swims around, calling or not, and attempts to mate with other frogs he encounters. If a receptive female is found, the male clasps her firmly to discourage other males from grabbing hold. Even so, females often become ensnared by two or more males (pictured on page 201). Wood Frogs typically lay their eggs communally to create large, intermingled masses in one or a few parts of the pond.

Voice: The advertisement call is a relatively soft, ducklike cackling: *ca-ha-ha-ac, ca-ha-ha-ac, ca-ha-ha-ac,* sometimes given in a rolling series.

Several males attempting to mount a lone female

58

Carpenter Frog

Rana virgatipes (Lithobates virgatipes) (1 5/8″–2 5/8″)

The Carpenter Frog gets its common name from its advertisement call, which sounds like carpenters driving nails—a distinctive calling card of coastal plain bogs and ponds. Because the species commonly resides in bogs that have abundant sphagnum moss, some people refer to it as the sphagnum frog.

Appearance: A Carpenter Frog can be identified by the two light longitudinal stripes along the sides of the back. Two additional light stripes occur lower on the sides, from the jaw to the groin. The upper lip is also light. The ground color ranges from brown or reddish brown to olive, and there are often dark spots between the stripes. Bronze and Green Frogs may look similar, but they have dorsolateral folds, which are absent in Carpenter Frogs. Bullfrogs lack the light longitudinal stripes. Immature Pig Frogs might cause confusion, but the webbing on their hind feet is much more extensive than that of Carpenter Frogs, in which the fourth toe is mostly free of webbing.

Range and Habitat: An Atlantic coastal plain species, ranging from the Pine Barrens region of New Jersey southward to southeastern Georgia and two counties in extreme northeastern Florida. Found in sphagnum bogs, acidic ponds, pocosins, and swamps, usually surrounded by sandy pine flatwoods.

Behavior: Carpenter Frogs are aquatic and live in the same habitats that they breed in from spring to late summer. Males are territorial and call from open water and floating vegetation—they may even sound off while perched on lily pads. Males are highly faithful to their territory and its immediate vicinity, and members of both sexes rarely leave their bog or swamp to wander overland.

Voice: The advertisement call is a distinctive series of sharp, doubled, rapping notes that sound like two carpenters hammering nails slightly out of sync: *c-tuck, c-tuck, c-tuck, c-tuck.* Males have large paired vocal sacs that extend out to the sides.

Florida Bog Frog

Rana okaloosae (Lithobates okaloosae) (1 3/8"– 1 15/16")

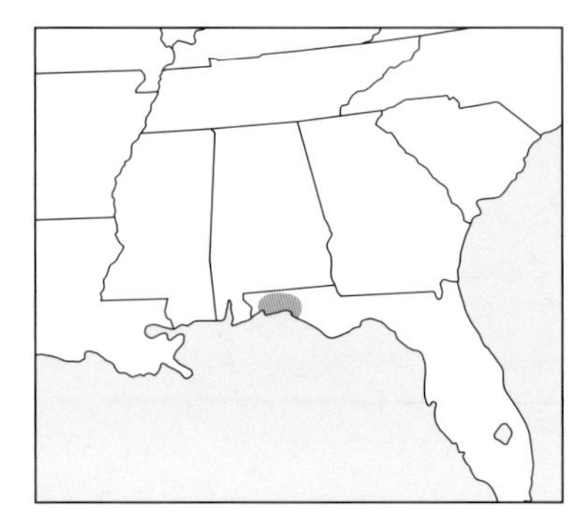

A close relative of the Green and Bronze Frogs, the rare Florida Bog Frog was first described in 1982 by Florida herpetologist Paul Moler and occurs only in a small area in the Florida Panhandle. Even in its favored habitat, this small frog might easily go unnoticed were it not for its unique rattling call.

Appearance: Less than two inches long when full grown, Florida Bog Frogs are the smallest of our native true frogs. Superficially, they look like small Bronze Frogs. The ground color varies from brown or brownish green to sphagnum green, there are often darkish spots on the back, and light spots may occur on the lower sides. The dorsolateral folds are light, the upper lip is greenish yellow, and the eardrum is brown. The undersides are dark with wormlike markings, contrasting with a yellow throat, and there is very little webbing on the hind feet (Bronze Frogs have extensive webbing).

Range and Habitat: Occurs in three stream drainages (Titi Creek, East Bay River, and the lower Yellow River) in three counties in the Florida Panhandle, with the majority of populations located within Eglin Air Force Base. Confined to the vicinity of acidic clear streams and seeps, as well as associated aquatic habitats such as boggy overflows dominated by sphagnum moss. Although populations appear to be in no immediate danger, these rare and extremely localized frogs are classified as a species of special concern in Florida.

Behavior: Florida Bog Frogs breed from April through August, with some calling as late as early September. Breeding activity occurs in the same habitat where they live year-round.

Voice: The advertisement call is a series of five to fifteen (or more) rattling, guttural notes given at a rate of four or five per second and dropping in volume at the end: *Currt-Currt-Currt-currt-currt.* Soft throaty notes or a staccato *PIT!* may also be given, especially in response to the calls of a nearby male.

River Frog

Rana heckscheri (Lithobates heckscheri) (3 1/4″– 6 1/8″)

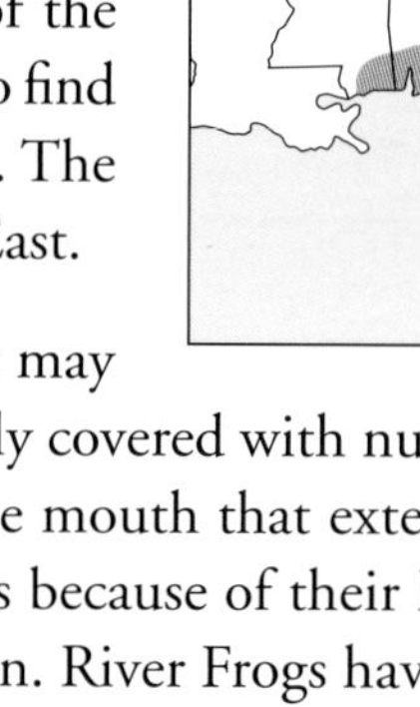

The River Frog is so named because it lives in the great river swamps of the southeastern United States. Though rarely heard, this large frog can be easy to find along the swampy edges of rivers and streams where it is locally common. The River Frog is not well studied and is one of the least known frogs of the East.

Appearance: The skin of the River Frog is rough, and the ground color may be gray, dark gray, or a dark brownish green. Lighter individuals are usually covered with numerous ragged black spots. There is a sprinkling of distinctive light spots along the sides of the mouth that extend to under the eardrums. River Frogs are likely to be confused with Bullfrogs and Pig Frogs because of their large size and lack of dorsolateral folds, but Bullfrogs and Pig Frogs are typically a brighter green. River Frogs have undersides that are dark gray and decorated with light spots and wavy lines, a pattern that is not typical of other large members of the genus.

Range and Habitat: A coastal plain species, the River Frog has a range thought to extend from southeastern North Carolina all the way to the Mississippi River, and in Florida south to the Ocklawaha River. But there are no recent records from North Carolina, and the species may no longer make it to the Mississippi. Prefers swampy areas along rivers and creeks, but also found in oxbow lakes and ponds.

Behavior: River Frogs breed from April to August, but calling is probably restricted to brief episodes within this period. Choruses are rarely heard. In fact, many experienced herpetologists have never heard calling individuals. In areas where River Frogs are most abundant, adults are easy to approach at night as they sit along the banks of streams. If caught, they often go limp, only to bound away when released.

Voice: The advertisement call is a robust belching snore lasting about two seconds and given at a leisurely rate. It sounds similar to the low-pitched snore of a Gopher Frog, although the latter is more drawn out and has a slower pulse rate.

Crawfish Frog

Rana areolata (Lithobates areolatus) (2 1/4"–3 5/8")

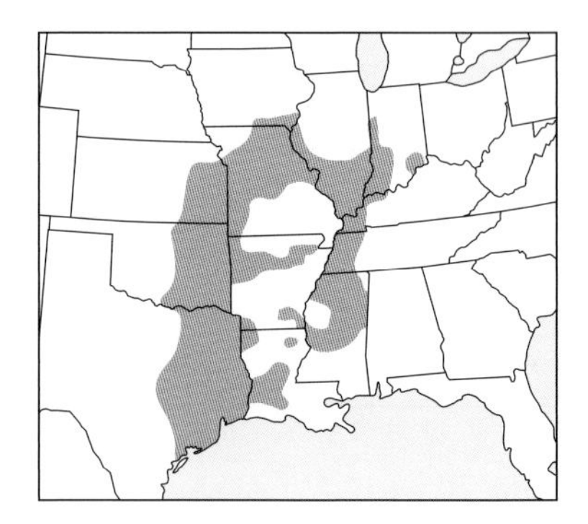

A close relative of the Gopher Frog, the Crawfish Frog gets its common name from its habit of sheltering by day in the abandoned burrows of crawfish. It also uses the burrows of other animals, and even artificial shelters such as sewers. Crawfish Frogs are listed as endangered or of special concern in several states, and local declines are mainly attributed to loss of habitat.

Appearance: The Crawfish Frog is chunky and toadlike, much like the Gopher Frog, but differs by having smooth skin (no warts) and dark spots that are edged in white on a variable ground color. The underside is white and unmarked, and males sometimes have a yellow wash on their dorsolateral ridges and concealed parts of their legs. Two subspecies are recognized. Southern Crawfish Frogs *(R. areolata areolata)* are smaller and have less prominent dorsolateral ridges and longer and narrower heads than do Northern Crawfish Frogs *(R. areolata circulosa).*

Range and Habitat: Has an odd arch-shaped distribution pattern. Ranges from eastern Texas northward along a narrow band all the way to southern Iowa, and then southward along a narrow band through Illinois and down to Mississippi. Largely absent from the interior of the arch, which includes southern Missouri and Arkansas (especially the Ozark Plateau) and parts of Louisiana. Found in a variety of habitats ranging from prairies, pastures, and meadows to pine forests, wet woodlands, and river floodplains.

Behavior: Breeds from late winter to early spring, depending on latitude and local weather patterns. As is the pattern with Gopher Frogs, heavy rains are required to initiate breeding, which is explosive and generally takes place over the course of several days to a week. Crawfish Frogs breed in farm ponds (without large fish), prairie wetlands, and shallow ditches. Males sometimes cluster together at a breeding site, favoring a particular spot along the shore or within a pool.

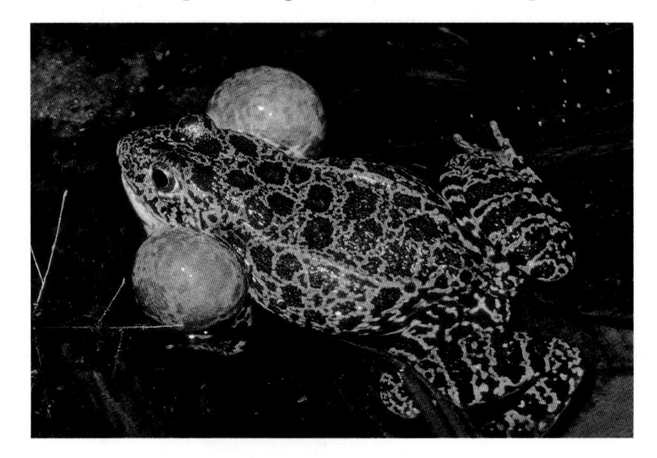

Voice: The advertisement call is a low-pitched, nasal, gagging snore that lasts about a second: *wwwwahhhhhh.* Neighboring males often call back and forth in close alternation. During aggressive encounters, a male gives a rapid, stuttered series of chucks that grow louder and end with an accented note: *c-c-c-c-c-c-c-cah!*

Southern Crawfish Frog

Gopher Frog and Dusky Gopher Frog

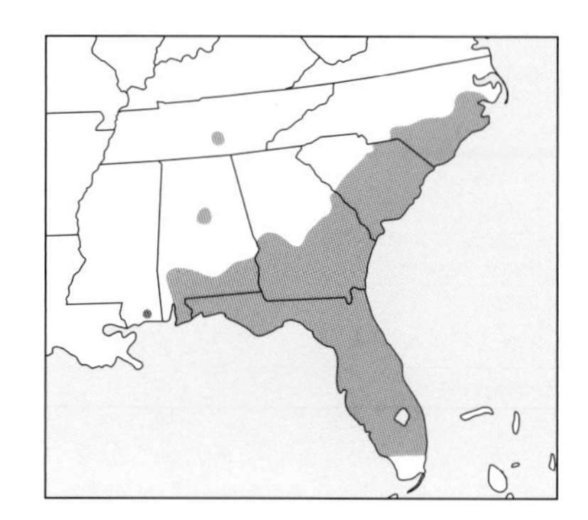

Rana capito (Lithobates capito) (2 3/4″–4 1/4″)
Rana sevosa (Lithobates sevosus) (2 1/2″–3 7/8″)

Stubby and toadlike in form, our two species of gopher frogs get their names from taking shelter in the burrows of the Gopher Tortoise, although they also use burrows made by other animals and other underground shelters. The Gopher Frog, *Rana capito,* has two recognized subspecies: the Carolina Gopher Frog *(R. capito capito)* and the Florida Gopher Frog *(R. capito aesopus).* The Dusky Gopher Frog, *Rana sevosa,* was formerly also considered a subpsecies of *Rana capito,* but recent genetic studies show that it is actually a full species. It is currently restricted to a single population in southern Mississippi (red dot on map). Gopher frogs are listed as endangered, threatened, or species of special concern wherever they occur.

Appearance: Gopher frogs have dorsolateral folds. The Carolina subspecies is gray to dark gray with roundish dark spots covering much of its upper parts, and numerous small warts crowded close together. In contrast, the Florida subspecies is usually lighter in color, ranging from cream to light brown. Its dark spots are irregular in size, and warts are much reduced or absent. The Dusky Gopher Frog is generally the darkest of the three (hence its common name) even though its color is changeable. Some individuals appear uniform black, while others are brown or gray with dark spots. Its undersides are heavily spotted, and warts are well developed.

Range and Habitat: The Carolina Gopher Frog ranges from North Carolina into Georgia, while the Florida Gopher Frog is found in Georgia, Florida, and Alabama. The Dusky Gopher Frog, which has been extirpated over most of its former range (from southwestern Alabama to Louisiana), now seems to be restricted to a small area in the De Soto National Forest in southern Mississippi. Gopher frogs prefer relatively dry areas, such as pine woods, turkey oak and pine sandhills, and old fields.

Behavior: Gopher frogs breed explosively in response to heavy rainfall, usually in late winter or early spring and sometimes in the autumn, although calling may occur in the southern part of the range during any month. Breeding takes place in temporary or semipermanent ponds, and also in ditches and swamps. Outside the breeding season, individuals take shelter underground but may emerge at night to feed near their burrow entrances.

Gopher Frog

Voice: The advertisement call of both species is a deep, rattling snore lasting about three seconds: *grrrraaawwwwwww.* Large breeding choruses produce an eerie, nonstop rumbling sound.

Light-colored Florida subspecies (Rana capito aesopus) *of the Gopher Frog*

Dusky Gopher Frog

Pickerel Frog

Rana palustris (Lithobates palustris) (1 3/4"–3 7/16")

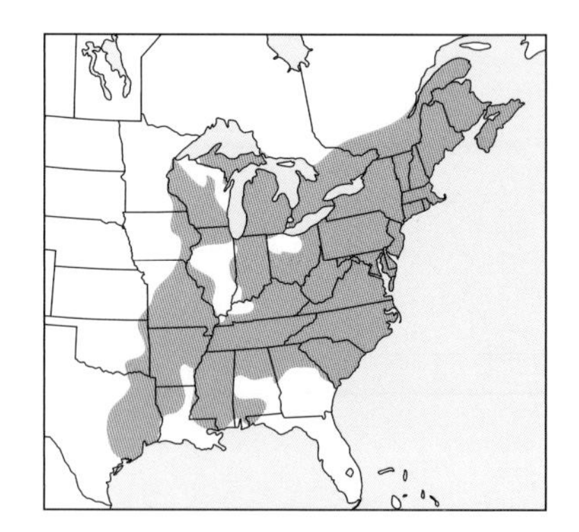

Adorned with squarish spots reminiscent of those found on pickerel fish, the Pickerel Frog has poisonous skin secretions that help protect it from predators. Herpetologists avoid transporting other species in the same bag as a Pickerel Frog because its secretions will kill the other frogs. The toxin is also capable of making a human sick, so you should wash your hands immediately after handling this frog.

Appearance: Pickerel Frogs are often confused with various kinds of leopard frogs found within their range. A distinguishing feature is that the chocolate brown spots on the back are typically squarish or rectangular and arranged in two (rarely three) parallel rows, bordered by light-colored dorsolateral folds. By contrast, the spots on the backs of leopard frogs are more circular in shape and more irregular in placement. Pickerel Frogs also have yellow or orange pigment on the concealed parts of their thighs in the groin. The ground color is generally light brown.

Range and Habitat: In Canada, Pickerel Frogs are found from southern Ontario to Nova Scotia. In the United States, they are common in the northeastern and Mid-Atlantic states, but range as far south as Georgia and Texas. In northern areas, Pickerel Frogs prefer habitats characterized by clear, cool water. Farther south, they are tolerant of the tannin-stained, warmer water of floodplain swamps. Usually found around the edges of wetlands, Pickerel Frogs frequently wander about on land and are often found in the twilight zone of caves.

Behavior: Pickerel Frogs are early-spring breeders, but they may call during warm spells in December in the South, and not until June at the northern limits of the range. In any one place, the breeding season is typically short, lasting only a few weeks. Males gather in woodland pools, ponds, springs, and other wetlands and frequently call from beneath the surface of the water.

Voice: The advertisement call is a soft grating snore lasting about two seconds (the snore of the Northern Leopard Frog lasts much longer and is followed by soft grunts). Garbled, throaty notes are occasionally given, along with a staccato *guck* that sounds much like the call of the Green Frog.

65

Northern Leopard Frog

Rana pipiens (Lithobates pipiens) (2″–4 3/8″)

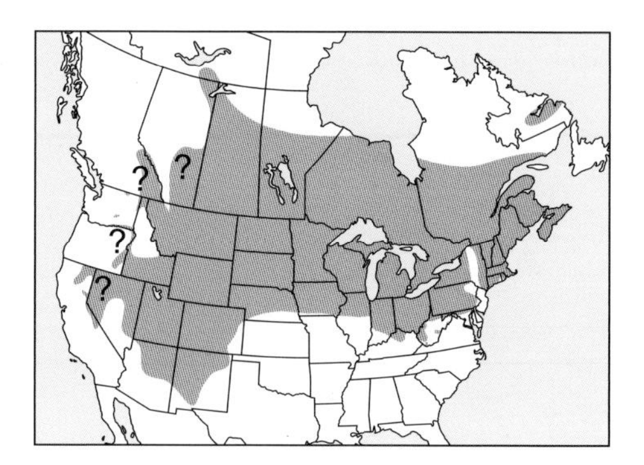

Common in areas with cold climates, Northern Leopard Frogs often frequent terrestrial habitats well away from their breeding ponds, especially wet meadows, which leads some people to refer to them as Meadow or Grass Frogs. The species is highly variable in appearance and best identified by its distribution and its distinctive advertisement call.

Appearance: The body color typically ranges from brown to green and is decorated with dark roundish spots with pale margins on the back and sides. Dorsolateral folds are prominent and complete. A light stripe is present on the upper jaw, and the eardrum usually lacks a light spot in its center. Variant color forms also occur, with some individuals having nearly black spots set against a light tan ground color (see photo below). In parts of Minnesota and North Dakota, one often finds individuals in which dark mottling obscures the spots (kandiyohi morph) or spots are lacking (burnsi morph)—these variations are pictured on page 219.

Range and Habitat: Wide-ranging across the northern third of the United States from Maine to Washington, and in southern Canada from Newfoundland to British Columbia. In the West, the species ranges southward through the Rocky Mountains and Great Basin into northern Nevada, northern Arizona, and New Mexico. Introduced populations occur in parts of California and extreme western Nevada. Population declines, severe in some areas, have occurred since the 1960s over much of the West. The causes are still being investigated, but disease, introduced predators, habitat destruction, and chemical pollution appear to be the primary culprits. Northern Leopard Frogs occupy a wide variety of habitats, including grasslands, woodlands, wet meadows, bogs, ditches, slow-flowing streams, and high mountain lakes.

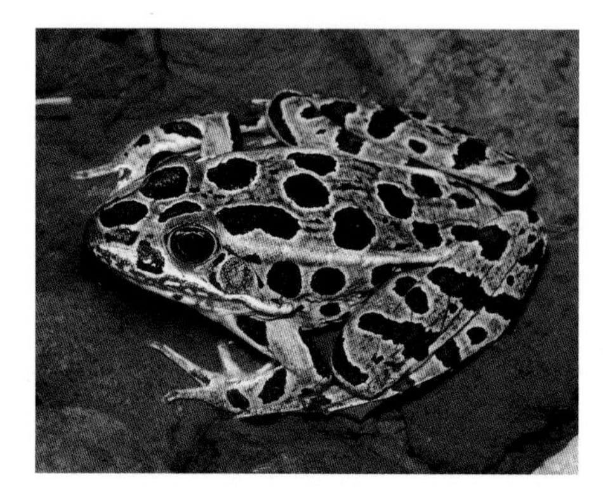

Behavior: Breeding occurs during warm periods from late winter to early spring, primarily in emergent vegetation along the edges of permanent streams, lakes, ponds, and pools. The breeding season lasts about a week in some areas and more than a month in others.

Voice: The advertisement call is a drawn-out rattling snore lasting three seconds or longer, usually followed by various soft grunts or chuckled notes. Snores start soft and grow louder before trailing off at the end. They sound quite different from the brief snores and chuckled outbursts of the various other leopard frogs found along the southern edge of the range.

Kandiyohi morph

Burnsi morph

Southern Leopard Frog

Rana sphenocephala (Lithobates sphenocephalus) (2"– 5")

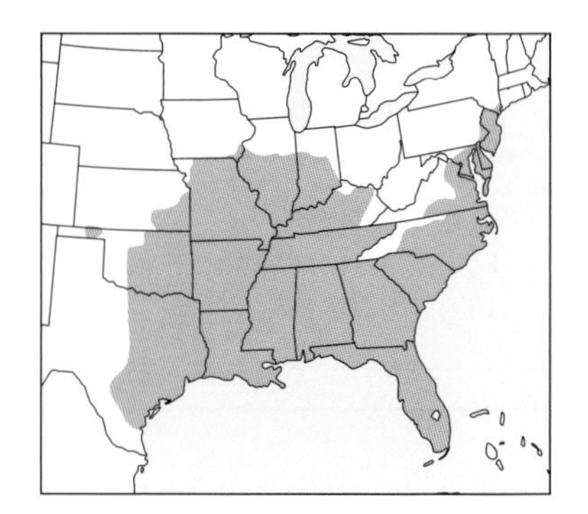

Wide-ranging in the Southeast, the lower Midwest, and the south-central states, the Southern Leopard Frog is a familiar and abundant spotted frog. They are found in wet areas of many kinds, and the raucous chuckles of males combine on warm spring nights to form a chattering cacophony that can be heard from hundreds of yards away.

Appearance: The Southern Leopard Frog can be distinguished from the three other leopard frogs that occur on the borders of its range (Northern, Plains, and Rio Grande Leopard Frogs) by a combination of features: a prominent light spot in the center of each eardrum, a relatively long and pointed head, reduced numbers of spots on the sides, dorsolateral folds that are usually continuous, and a distinctive advertisement call. Ground color ranges from brown to gray or green or some combination of these. Roundish dark spots may be numerous to sparse.

Range and Habitat: The eastern edge of the range extends from New York to the Florida peninsula. The western edge arcs from Missouri across southeast Kansas and through the eastern half of Oklahoma and Texas. The northern edge slices through Illinois and Indiana before looping south around the Appalachian Mountains, in which this frog is notably absent. Southern Leopard Frogs frequent a wide variety of shallow freshwater and even brackish-water habitats, as well as moist, thickly vegetated areas well away from water.

Behavior: Breeding occurs in the spring in the northern part of its range and whenever there are heavy rains in southern areas. Agile and wary, Southern Leopard Frogs are difficult to catch. When captured, an individual of either sex may give an ear-piercing scream that might startle a predator, causing it to drop the frog.

Voice: The advertisement call, which is repeated several times in rapid succession, consists of a series of five to ten (or more) guttural chucking notes that are delivered as a rapid stutter: *chu-h-h-h-huck, chu-h-h-h-huck, chu-h-h-h-huck,* often followed by grunting or scraping sounds. At low temperatures, chucks are given at a slow rate and may sound similar to the slow-paced calls of the Plains Leopard Frog. At warmer temperatures, chucks are higher in pitch and are delivered so rapidly that they sound like a rattle.

Plains Leopard Frog

Rana blairi (Lithobates blairi) (2″–4 3/8″)

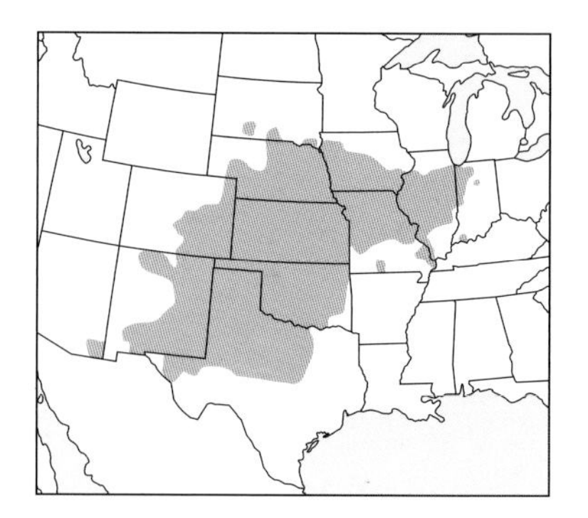

The Plains Leopard Frog is associated with plains and prairies throughout most of its range. The scientific name *blairi* pays tribute to biologist W. Frank Blair (1912–1984), a pioneer in the study of frog vocalizations and their evolutionary significance. Although mostly aquatic, this frog may be found considerable distances from standing water during rainy periods.

Appearance: The dorsal color is generally brown to tan but is occasionally dark green. There are usually numerous roundish dark brown spots on the back that are not encircled with lighter pigment, and a distinct white line along the upper lip. The eardrum typically has a light spot in its center. There are prominent dorsolateral folds that become broken or discontinuous near the thigh. Many individuals have yellow pigment in the groin and on the underside of the thigh.

Range and Habitat: The Plains Leopard Frog is an inhabitant of the central plains from northern Texas to Kansas and southeastern South Dakota. To the west, it ranges into Colorado and New Mexico, with isolated populations in New Mexico and southeastern Arizona that have greatly declined in recent years. To the northeast, the range extends eastward along a "prairie peninsula" through parts of Iowa and Missouri into Illinois and northwestern Indiana, and along the Mississippi River as far south as Arkansas. Although they prefer plains, prairies, and open river floodplains, Plains Leopard Frogs sometimes wander into upland habitats. They are usually found in and around permanent and semipermanent bodies of water, including ponds, streams, ditches, and cattle tanks.

Behavior: Over most of their range, Plains Leopard Frogs breed in early spring after the first warm rains, but they may breed in any month of the year in warm and arid southern regions where rainfall is unpredictable. Escaping individuals often leap away from, rather than toward, the water.

Voice: The advertisement call consists of two to four throaty, chucking notes, repeated several times in rapid succession: *chu-hu-huck, chu-hu-huck, chu-hu-huck* or *hi-hi-hip, hi-hi-hip, hi-hi-hip*. Each outburst of chucks rises slightly in pitch and ends with an accent. A soft grunt may terminate a series. Males also produce squeaking notes reminiscent of the sounds made by rubbing one's finger across the surface of an inflated balloon.

Rio Grande Leopard Frog

Rana berlandieri (Lithobates berlandieri) (2 1/4"–4 1/2")

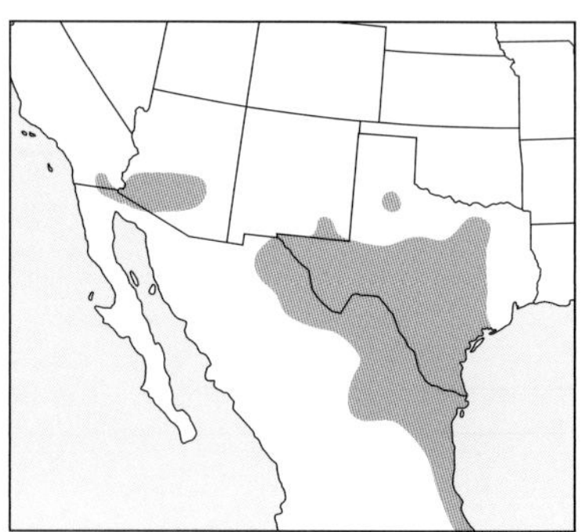

Largely confined to arid lands of Texas, New Mexico, Arizona, and adjacent Mexico, Rio Grande Leopard Frogs are often paler in color than other leopard frogs. They are usually found in or near aquatic habitats and may sometimes be observed sitting or floating in patches of emergent vegetation or on the banks of ponds or streams.

Appearance: The dorsal ground color is typically a pallid tan, gray, or beige, often with green on the head, but some individuals are mostly green or olive. The eardrum may or may not have a light central spot; there are usually no spots on the head in front of the eyes; and the dorsolateral folds are interrupted and noticeably inset toward the rear. The light jaw line becomes indistinct in front of the eyes. There is a reticulated or weblike pattern of dark markings on the rear of the thighs, and adult males have prominent vocal sacs.

Range and Habitat: Found primarily in south and southwestern Texas and northeastern Mexico but ranges as far north as the Dallas area and also just into southeastern New Mexico. Introduced and expanding populations occur in riverine and agricultural lands in southwestern Arizona and southeastern California (and adjacent areas of Mexico). Often found in permanent and semipermanent streams, but may also occur in rivers, ponds, springs, ditches, canals, and cattle tanks.

Behavior: Breeding in the Rio Grande Leopard Frogs occurs during the warmer months and may or may not be associated with rainfall. Adults remain active year-round in warmer areas. They are primarily aquatic but may be found a mile or more from water during and after summer storms.

Voice: The advertisement call is a loud, guttural, rattling snore lasting about half a second, given singly or in groups of two or three, and often followed by soft garbled notes: *rawwwwwwk, rawwwwwwk, rawwwwwwk . . . gurglela . . . gurglela.* A staccato *chuck!* may occasionally also be given. These calls can easily be confused with those of the Southern Leopard Frog, which may share habitats with the Rio Grande Leopard Frog in eastern Texas.

Relict Leopard Frog

Rana onca (Lithobates onca) (1 3/4"–3 1/2")

The Relict Leopard Frog has suffered severe declines and now occurs in only a few sites in southern Nevada and perhaps in adjacent Arizona. This species was considered extinct from 1984 until its rediscovery in 1991. Although it is closely related to the Lowland Leopard Frog, genetic studies confirm that Relict Leopard Frogs are distinct.

Appearance: Similar in appearance to the Lowland Leopard Frog in that both are relatively small leopard frogs with dorsolateral folds that are usually broken and inset toward the rear. The Relict's hind legs are proportionately shorter than the Lowland Leopard Frog's, and spots on the dorsal surface tend to be reduced or obscure toward the front of the animal. Coloration varies from light brown or tan to dark olive or charcoal, with dark spots. The ventral surface is whitish with some faint markings, and the undersides of the hind limbs are yellow or orange-yellow.

Range and Habitat: The historic range of the species included wetlands along several river systems in southwestern Utah, northwestern Arizona, and southern Nevada. Introduced Bullfrogs, crayfish, and sport fishes, along with habitat alteration due to agriculture and water management, all contributed to declines. Now the species is confined to springs, spring-fed wetlands, and short stream segments that drain into the Overton arm of Lake Mead and to the Colorado River in Black Canyon below Hoover Dam. These few remaining natural populations have recently been augmented by reintroductions to other aquatic habitats in the area.

Behavior: Judging from observations of eggs and tadpoles, Relict Leopard Frogs breed from February through April, and also in November, in pools and in slow-moving areas of streams. Males are rarely heard and may do most of their calling underwater. Given these frogs' perilous status, biologists are working hard to gather detailed information about their behavior and natural history.

Voice: The advertisement call is a series of five to ten spirited chuckles given in rapid succession and sometimes followed by grating notes: *chuckleluck-chuckleluck-chuckleluck-chuckleluck.* The sounds are virtually identical to the call of the Lowland Leopard Frog but easily distinguished from the more prolonged snore of the Northern Leopard Frog.

Chiricahua Leopard Frog

Rana chiricahuensis (Lithobates chiricahuensis) (2″–5 2/5″)

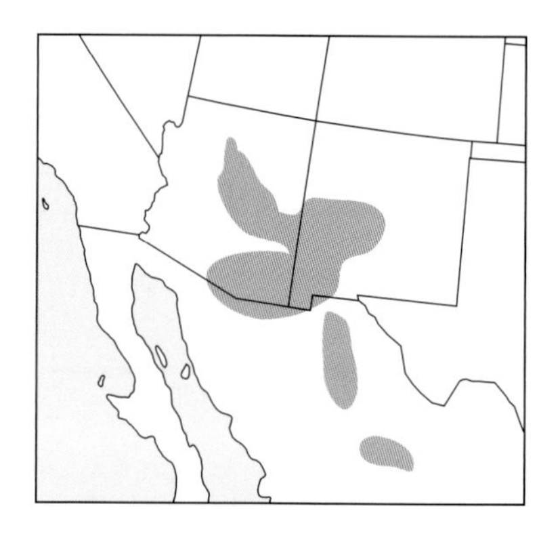

Confined to scattered localities in the mountains and high valleys of Arizona, New Mexico, and adjacent Mexico, the Chiricahua Leopard Frog's common name is derived from the fact that its type locality (where the specimen used to define the species was collected) is in the Chiricahua Mountains of southeastern Arizona. The Chiricahua Leopard Frog has experienced significant declines in recent years and is listed by the U.S. government as threatened.

Appearance: Chiricahua Leopard Frogs are characterized by their relatively rough skin, slightly upturned eyes, and dorsolateral folds (prominent in southern populations) that are broken and inset near the rear. The ground color is green or brown, and there are more spots of relatively small size in comparison with other leopard frogs. The area above the upper jaw is usually green, and the light stripe along the upper lip is indistinct, especially in front of the eye.

Range and Habitat: The range of the Chiricahua Leopard Frog is roughly divided into two sections. A northern series of populations occurs in the mountains along the southern edge of the Colorado Plateau, known as the Mogollon Rim, which runs from central and eastern Arizona to west-central New Mexico. A more southerly series of populations is found in the mountains and valleys of southeastern Arizona, southwestern New Mexico, and adjacent areas in Mexico. Arid landscapes separate these two population groups, and they may be described as separate species in the future. A few isolated populations occur in the Huachuca Mountains in southeastern Arizona. Individuals at these locations call almost exclusively while underwater and are considered by some to be a separate species, the Ramsey Canyon Leopard Frog *(R. subaquavocalis).* Previously found in a wide variety of permanent and near-permanent bodies of water, Chiricahua Leopard Frogs are now restricted to headwater springs, cattle tanks, and other sites lacking introduced predators such as Bullfrogs, crayfish, and sport fishes.

Behavior: Breeding occurs from February through October, whenever water temperatures are at or above about 57°F. Males may call from above or below the water's surface.

Voice: The advertisement call is a grating snore that lasts about a second and often ends with an accented *chuck!* Staccato chucks are sometimes given singly.

Lowland Leopard Frog

Rana yavapaiensis (Lithobates yavapaiensis) (1 4/5"–3 2/5")

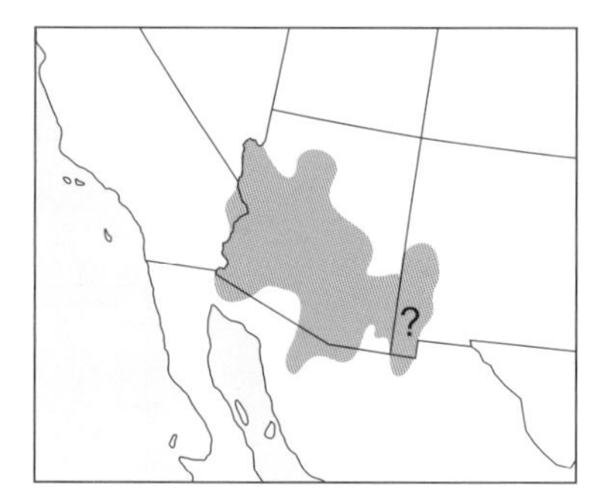

Similar in appearance to the Relict and Rio Grande Leopard Frogs but genetically distinct, the Lowland Leopard Frog has a limited range in riparian areas of the Southwest. While populations appear stable in central Arizona, there have been severe declines and local extirpations in many other areas for unknown reasons. Chytrid fungus has likely played a role, but populations may also have been eliminated by introduced predators such as Bullfrogs, crayfish, and sport fishes.

Appearance: Lowland Leopard Frogs are relatively small leopard frogs with dorsolateral folds that are broken or inset toward the rear. The head in front of the eyes is usually free of spots, and there is a distinctive dark brown reticulated pattern on the rear of the thigh. The ground color ranges from tan to brown, although some are green, particularly on the head. The Lowland Leopard Frog does not have a complete, light stripe on the upper lip. Adults males lack vocal sacs.

Range and Habitat: Found from northwestern Arizona through central and southeastern Arizona to southwestern New Mexico (where it may be extirpated) and adjacent Sonora. It was once found in extreme southeastern California and was also found in big rivers such as the Colorado and Gila, where it is now absent. Habitats range from desert scrubland to conifer and pine-oak woodlands. Occupies a wide variety of permanent and semipermanent streams, ponds, springs, ditches, and cattle tanks.

Behavior: Lowland Leopard Frogs remain active most of the year at warm desert locations. They breed from January through April and then again in late summer and early fall. Hybridization with the Chiricahua Leopard Frog has been documented in areas of overlap.

Voice: The advertisement call is a brief chuckle composed of about three to six pulses, repeated in rapid succession: *chucklelah-chucklelah-chucklelah-chucklelah.* Another call is a grating *rrraack* that is reminiscent of the sound made by rubbing one's finger across a balloon.

Tarahumara Frog

Rana tarahumarae (Lithobates tarahumarae) (2½″–4½″)

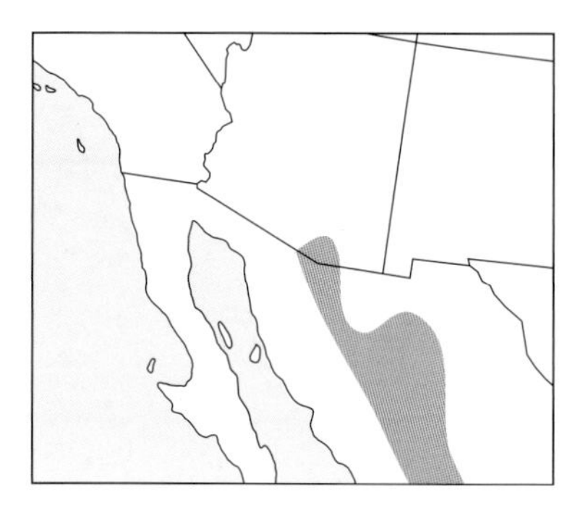

Tarahumara Frogs formerly occurred in isolated populations in extreme southern Arizona and south into northern Mexico, but the Arizona populations disappeared in the 1970s and early 1980s. The species was experimentally reintroduced from Mexico to a historic locality in Arizona in 2004. As of this writing, the introduced frogs are persisting in a single canyon, but the population is threatened by the chytrid fungus, and habitats have been degraded by sedimentation that occurred following a wildfire.

Appearance: Tarahumara Frogs are medium to large, rough-skinned frogs with dark spots on a ground color that varies from olive to dark brown. Unlike Bullfrogs, which they resemble superficially, Tarahumara Frogs have distinctly barred hind legs, cream or dusky throats (mottled in the Bullfrog), and poorly defined external eardrums. Dorsolateral folds are weak or absent.

Range and Habitat: Natural populations are currently found in the Sierra Madre Occidental and associated sky islands of Mexico. In the United States, a single reintroduced population persists in a remote canyon in the Santa Rita Mountains, where they originally lived. Frogs also occur at several experimental breeding sites elsewhere in the state, well outside of their historic range. In Arizona, Tarahumara Frogs prefer permanent streams in rocky canyons located in oak and pine-oak woodlands. They live in plunge pools and rocky, water-filled potholes that are found primarily in mountains but that also occur in semidesert grasslands. Tarahumara Frogs are highly aquatic and are rarely found more than a few feet from water.

Behavior: Tarahumara Frogs breed primarily in the spring but may also breed during monsoon rains in late summer. Calling occurs in both seasons. Males don't have vocal sacs but are able to produce a variety of sounds.

Voice: The advertisement call appears to be a grating snore, lasting about a second and repeated often. Grunts, squawks, expressive peeps, and whines may also be given. None of the calls is very loud.

Cascades Frog

Rana cascadae (1 3/4″–3″)

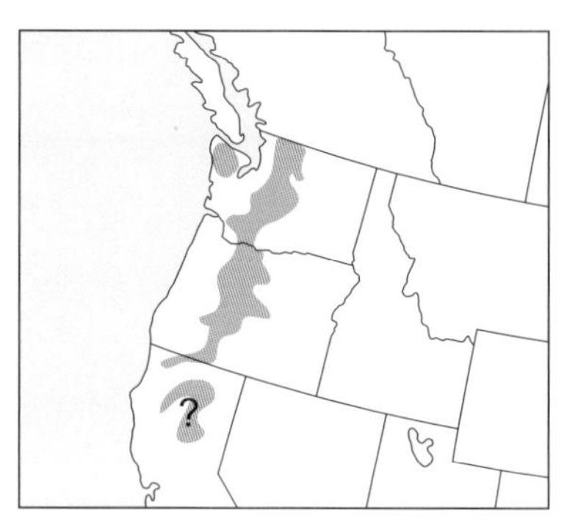

Thriving up to the timberline in the Cascade Mountains, the Cascades Frog is a common resident of high-elevation aquatic habitats, where individuals may be found in streams or sitting exposed at the water's edge. The species is doing well throughout most of its range but has experienced major declines in southern locations.

Appearance: Cascades Frogs are typically olive, brown, reddish brown, or tan in color, and they usually have a scattering of jet black, even-edged spots on their backs. Dark spots or bands also occur on the legs, and most individuals have dark masks. The undersides of the body and legs are yellow, flesh colored, or orange-yellow, and the groin may have indistinct dark mottling. Complete dorsolateral folds are prominent.

Range and Habitat: Found most commonly in the Cascade Mountains of Washington, Oregon, and northern California. Isolated populations occur in the Olympic Mountains of Washington. Cascade Frogs were once extremely abundant in the Lassen Volcanic National Park and surrounding areas in northern California, but for unknown reasons the frogs have almost completely disappeared from those areas. Disease is a likely cause of declines, with introduced fish and pesticides as possible contributing factors. Cascades Frogs inhabit moist meadows and other open wetlands, small streams, pools, ponds, and lakes.

Behavior: Cascade Frogs are often slow moving and easily approached. When alarmed, they may jump into the water, only to reappear at the original location a minute or two later. They breed soon after snowmelt, from March to August, depending on the elevation and location.

Voice: The advertisement call is a rapid chuckle, sometimes ending with a staccato grunt or growl: *k-k-k-k-k-k-kuk!* The tempo of the call slows considerably when it is cold. Calls are given both in the air and underwater and can be heard only over short distances.

Northern Red-legged Frog

Rana aurora (1¾″–3″)

The Northern Red-legged Frog is named for the red on the undersurfaces of its long hind limbs. It is an excellent leaper and uses this ability to avoid predators. It is similar in appearance to but smaller than the California Red-legged Frog, and the two species were long considered to be subspecies. Recent genetic studies have shown, however, that not only are they two separate species, but the Northern Red-legged Frog's closest relative is the Cascades Frog, not the California Red-legged Frog.

Appearance: The ground color may be brown, rusty red, olive, or gray, and scattered small dark spots or blotches may exist on the head, back, and sides. Dark bands usually occur on the legs, and a dark mask may also be present. A light jaw stripe extends to the shoulder, and on the lower back there are prominent dorsolateral folds that have a slight offset.

Range and Habitat: Found from extreme southeastern Alaska to northern California in the lowlands and foothills between the coast and inland mountains. Occurs in and near ponds and streams in a variety of habitats, where individuals take refuge in dense aquatic and semiaquatic vegetation. Outside the breeding season, they disperse widely and may wander far from their favored breeding spots, where they frequent damp and wet upland vegetation.

Behavior: The breeding season lasts only a few days to a few weeks and may occur from November to April, depending on local climate and elevation. Breeding takes place in vegetated shallows along the edges of pools, ponds, and other wetlands.

Voice: The advertisement call is a chuckle composed of about four to seven notes, with emphasis at the end: *huh-huh-huh-huh-huh!* Chuckles occasionally end with a growl or groan like that of the California Red-legged Frog. These calls are weak and are usually given underwater, so choruses are easily missed. Some males in the southern populations have rudimentary vocal sacs, but most males of this species do not.

California Red-legged Frog

Rana draytonii (1 3/4"–5 1/4")

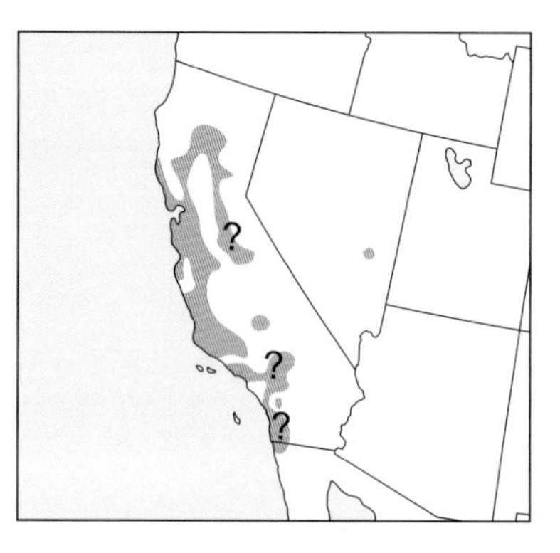

The California Red-legged Frog was very likely the subject of Mark Twain's short story "The Celebrated Jumping Frog of Calaveras County." It is the largest native frog in the West, reaching lengths of over five inches. The species is now threatened, having disappeared from no less than three-quarters of its former range because of urbanization, predation by nonnative Bullfrogs, and possibly pesticides and disease.

Appearance: As in the Northern Red-legged Frog, the undersides of the hind limbs are often red, and there are dark spots or blotches on a lighter ground color that ranges from reddish brown to olive or gray. Both species have dark bands on the legs, a dark mask above a light jaw stripe, and prominent dorsolateral folds. The two species differ in that the larger California Red-legged Frogs have more numerous spots (often with light centers), rougher skin, proportionally shorter legs, and smaller eyes.

Range and Habitat: The California Red-legged Frog now occurs mostly along the central California coast and coastal mountains, although a very few scattered populations remain in the Sierra Nevada foothills and in northern Baja California. There are several introduced populations in Nevada, but their status is uncertain. This frog is most common in lowlands and foothills, where it frequents ponds, streams, and other wetlands in coastal drainages. Individuals can often be found in dense vegetation near deep pools.

Behavior: Breeding lasts only a week or two and may occur from December through April, usually after heavy rains. Insects are an important food item, as they are for most frogs, but California Red-legged Frogs regularly eat small mammals and other frogs.

Voice: The quiet advertisement call is a series of four to seven low chucks, often ending with a scraping growl or groan: *huh-huh-huh-huh-huh . . . rawk.* Males have small, paired vocals sacs and call both day and night, sometimes from underwater. Their calls are typically weak, and choruses are easily missed.

Oregon Spotted Frog and Columbia Spotted Frog

Rana pretiosa (1 3/4"–4") and Rana luteiventris (1 3/4"–4")

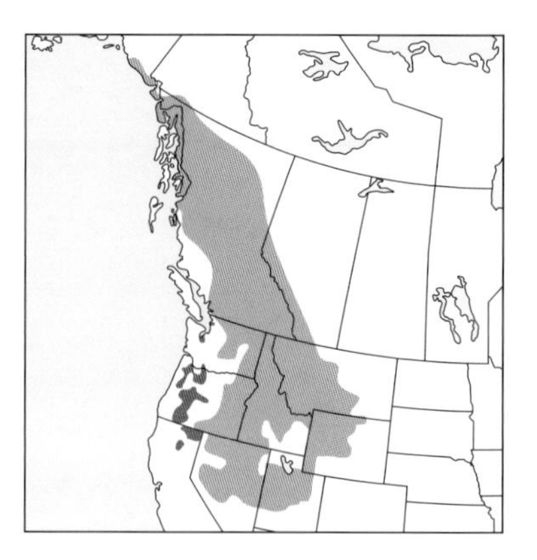

The Oregon and Columbia Spotted Frogs are named for both their ranges in the Pacific Northwest and for the prominent ragged-edged dark spots (usually with light centers) that are scattered on their backs. Formerly considered to be geographic races of one species, the two cannot be easily distinguished by appearance and are best identified by where they occur.

Appearance: These highly aquatic frogs have extensive webbing between the toes of the hind feet. The dorsal ground color varies from brown and tan to rust or red. Each species has a wash of bright color on its underside that broadens with age until it covers most of the undersurface except the throat. Juveniles of both species have cream-colored undersides, but the adult Oregon Spotted Frog has rust or orange-red underparts while those of the adult Columbia Spotted Frog are yellow to orange.

Range and Habitat: The Columbia Spotted Frog (orange on map) is wide-ranging in the Northwest, where it occurs throughout most of British Columbia, south through most of eastern Washington and Oregon, and east to isolated areas in Wyoming, Utah, and Nevada. It is found in diverse habitats, ranging from boreal ponds in the north to mountaintop wetlands and wet springs in southern arid regions. In contrast, the Oregon Spotted Frog (red on map) has a very restricted range, from extreme southwestern British Columbia southward through western Oregon and Washington to northeastern California. It is much more of a habitat specialist, preferring large marshes and the edges of lakes where the water is warm and there is a lot of low vegetation.

Oregon Spotted Frog

Behavior: Both species are typically early-season breeders, but breeding dates vary greatly, from February to the summer months, because of the enormous differences in altitude and weather conditions within their ranges. Breeding usually occurs in seasonal flooded shallow areas adjacent to lakes, springs, ponds, and marshes.

Voice: The advertisement call of both species is an extended series of hollow clicks, chucks, or pops, given in a relatively steady series that usually lasts ten seconds or more: *cok-cok-cok-cok-cok-cok-cok-cok-cok-cok.*

Columbia Spotted Frog (above and to the right)

Foothill Yellow-legged Frog

Rana boylii (1½"–3⅕")

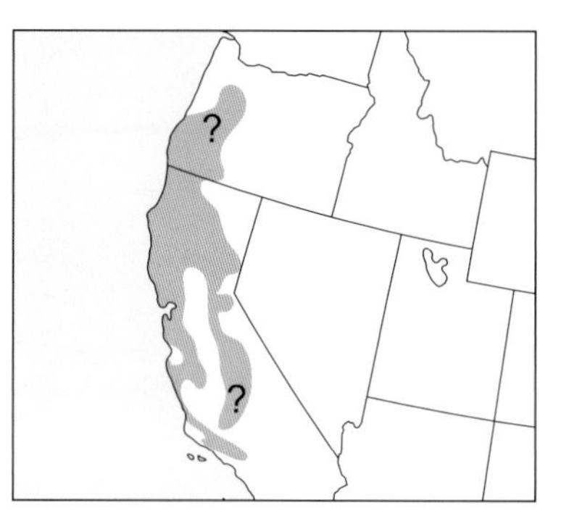

The Foothill Yellow-legged Frog is a stream-dwelling species found mostly in foothills from sea level up to about 6,500 feet. It has disappeared or declined over large parts of its range, possibly because of pesticides, introductions of nonnative Bullfrogs, stream-flow alteration due to dams, and disease.

Appearance: A prominent identification mark is the yellow wash found on the undersides of the hind legs and lower abdomen. The color and spotting of its granular back often matches that of the rocks in its surroundings and is usually some shade of gray, olive, brown, or occasionally red. Dorsolateral folds are poorly defined, and a pale triangular patch often occurs between the eyes and the snout.

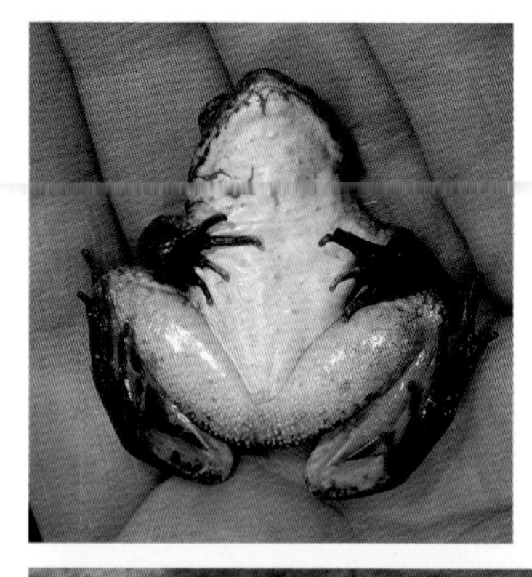

Range and Habitat: Currently found in Oregon west of the crest of the Cascades and southward into central California in the foothills of the Coast Range and along the western slopes of the Sierra Nevada Mountains. Robust populations still occur in the North Coast Range in California and in the Umpqua River system in Oregon, but declines in other areas are alarming. The species is almost extinct in the Willamette and upper Rogue River systems in Oregon, where it used to be abundant. Severe declines have also occurred in the southern Sierra Nevada Mountains and coastal areas in southern California. Primarily found in slow-moving rocky streams, often in and near riffles in sunlit areas.

Behavior: Adults are active during the day and are usually found next to water. The breeding season is from March to mid-June but varies in duration and timing, depending on local climate. In streams with high flow due to winter runoff, frogs wait for water levels and velocities to drop before breeding.

Voice: Foothill Yellow-legged Frogs make a variety of sounds. One call is a nasal, grating squawk. It usually lasts about half a second but may be longer and garbled at the end. Also gives a rapid-fire chuckle and staccato squeaks and grunts. Males typically call while underwater, but weak in-air calls are occasionally heard.

79–80

Sierra Madre Yellow-legged Frog and Sierra Nevada Yellow-legged Frog

Rana muscosa (1 3/5"–3 1/2") and *Rana sierrae* (1 3/5"–3 1/2")

Noted for their garliclike odor when handled, the Sierra Madre and Sierra Nevada Yellow-legged Frogs were formerly considered one species. They inhabit mountainous areas and historically have been found up to the highest elevation habitats that are free of snow and ice during the summer months. Both have experienced severe population declines in recent years.

Appearance: Both species range in color from olive to brown, yellow-brown, or red-brown and have dark spots or netlike patterns of dark pigment on the backs. The undersides of the hind limbs and often the belly are tinged with yellow or orange. Dorsolateral folds are sometimes evident. The legs of the Sierra Madre Yellow-legged Frog are proportionally longer than those of the Sierra Nevada Yellow-legged Frog, and there are subtle differences in their calls (see below).

Range and Habitat: The Sierra Madre Yellow-legged Frog of southern California and the southern Sierra Nevada is now almost extinct, with only a few small populations remaining (orange on map). The causes of its declines are unknown. The more northerly Sierra Nevada Yellow-legged Frog occurs at high elevations in the Sierra Nevada Range of California, where it can be found along streams, rivers, and mountain lakes, and in flooded meadows (red on map). This frog is also experiencing a severe decline and is now found in less than 5 percent of its historic localities. The main cause of the decline is the chytrid fungus, but the introduction of nonnative trout is responsible in some areas, and pesticide drift may be a contributing factor.

Sierra Nevada Yellow-legged Frog

Behavior: Both species call as soon as snow and ice melt and stream levels subside. This can occur as early as March at lower elevations and as late as June at high elevations. Most calling takes place underwater, hence the calls are seldom heard.

Voice: The advertisement call of the Sierra Nevada Yellow-legged Frog is a series of raspy, scraping sounds, sometimes ending with a louder, accented note, and it may also give stuttered notes. Calling sounds strained, as if the frog is struggling to make the call. The call of the Sierra Madre Yellow-legged Frog sounds similar but is less strained than that of the northern species.

Sierra Madre Yellow-legged Frog (above and to the right)

North American Spadefoots — Family Scaphiopodidae

Spadefoots get their name from the spade-shaped tubercles on their hind feet; these allow them to burrow backward into soil and sand (see photo and diagram). These chunky, toadlike frogs are nocturnal, have vertical pupils, and are well adapted to arid environments.

The family Scaphiopodidae has been recently split from the family Pelobatidae, which included spadefoots from Europe, western Asia, and northern Africa. Referred to as North American Spadefoots, seven species exist, and they are found in North America only. While there are good reasons for placing our spadefoots in their own family, there is some debate about whether they should be further divided into two genera, as we classify them in this book: *Scaphiopus* (southern spadefoots) and *Spea* (western spadefoots). Some authorities believe there should be only a single genus, *Scaphiopus*.

Spadefoots are mostly explosive breeders, requiring very heavy rainfall to trigger choruses and egg-laying. Breeding typically lasts only one or two nights. Amplexus is inguinal, with the male grasping the female around her waist. Spadefoot tadpoles develop into adults in a very short period of time, usually within two to three weeks. The rapid development is undoubtedly an adaptation to the transient nature of shallow breeding pools, and breeding may be skipped entirely during dry years. Adults may also estivate in burrows in response to drought, covering themselves with skin secretions to prevent water loss through evaporation.

Eastern Spadefoot — close-up of spade

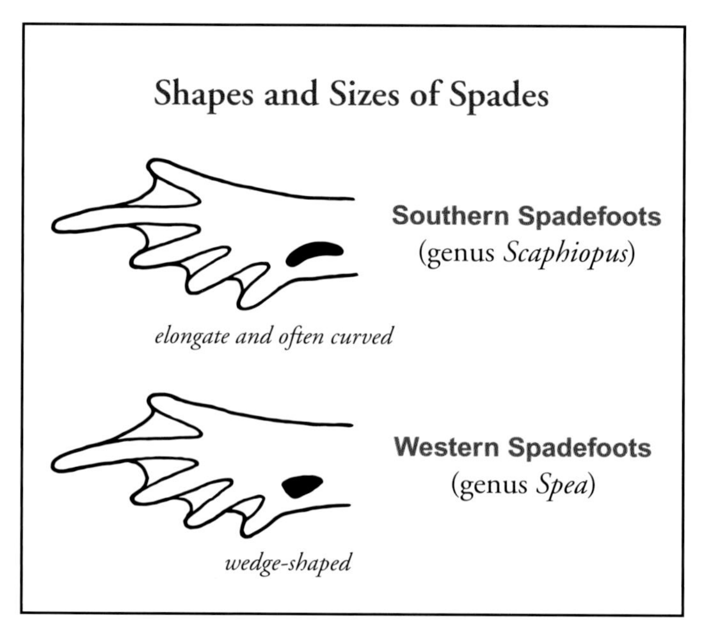

Eastern Spadefoot

Scaphiopus holbrookii (1 3/4"–2 7/8")

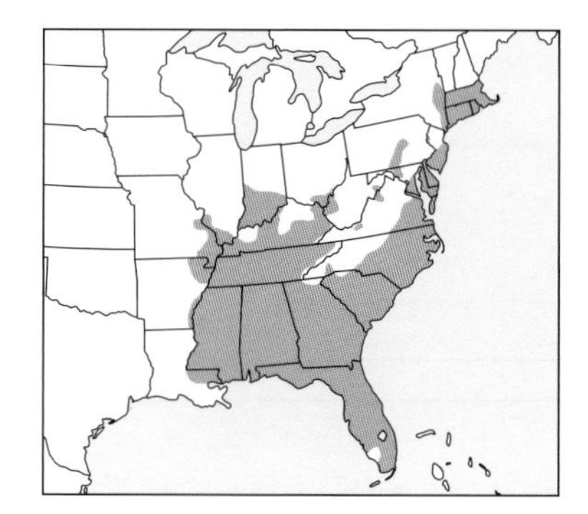

The only spadefoot found east of the Mississippi River, the Eastern Spadefoot is primarily nocturnal and is not often seen except after heavy rainfall, when its nasal croaking calls bring attention to its presence in breeding pools. Otherwise, these frogs spend most of their time underground, wandering short distances from their burrows on humid nights.

Appearance: Like other members of the spadefoot family, the Eastern Spadefoot has vertical pupils. It is distinctly marked on its back with a yellow, hourglass-like pattern set against a dark brown or blackish background, and there may be a scattering of small pink or orange tubercles. Yellowish spots and blotches also occur along the sides. The space between the eyes on the top of the head is flat, and the black spade on the hind foot is elongate and often curved (page 254).

Range and Habitat: Primarily southern in distribution, ranging throughout the Southeast, northward into Illinois, Indiana, and Ohio, and along the coast all the way to Massachusetts. Prefers dry forested areas, especially pinewoods and pine-oak woods with well-drained sandy soils.

Behavior: The Eastern Spadefoot is an explosive breeder that emerges and migrates to temporary ponds, pools, and ditches after heavy rains anytime from late winter to autumn (or in any month of the year in Florida). Mating and egg-laying usually take place in a single night. The eggs and tadpoles develop rapidly. These behaviors and developmental characteristics are also found among western spadefoots, which typically live in much more arid areas where rainfall is spotty and less predictable. This suggests that the Eastern Spadefoot evolved from a common ancestor in the West.

Voice: The advertisement call is an explosive, nasal utterance, usually down-slurred and sounding like a person gagging: *errrrrrrah!* Calls are repeated every five to ten seconds. The male has a large vocal pouch. When calling, he throws his head upward and backward out of the water, and the nictitating membranes (inner eyelids) cover his half-closed eyes.

Hurter's Spadefoot

Scaphiopus hurterii (1 3/4"– 3 1/4")

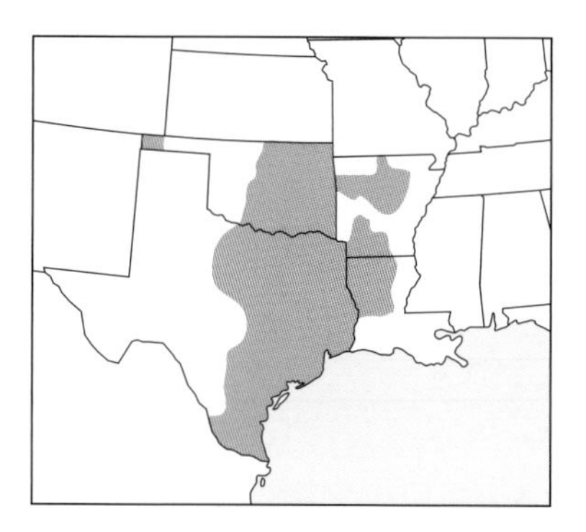

Restricted in range to the south-central states, Hurter's Spadefoot was formerly considered a subspecies of the Eastern Spadefoot, and some experts still consider it so. It is named in honor of Julius Hurter (1842–1917), a midwestern herpetologist. Hurter's Spadefoot is secretive and fossorial in habits, and its natural history is similar in most respects to that of the Eastern Spadefoot.

Appearance: Hurter's Spadefoot is rather plain looking, with a ground color that is a varying mixture of yellow and brown. A light hourglass-like pattern may occur on the back but is often missing. There are numerous small, rounded warts. Unlike its eastern relative, Hurter's Spadefoot has a raised area (boss) between and slightly behind the eyes. The spade is elongate and often curved (page 254).

Range and Habitat: Found in the eastern parts of Texas and Oklahoma and adjoining areas in Louisiana and Arkansas; its range does not overlap with that of the Eastern Spadefoot. Lives in a variety of habitats from sandy woodlands and grasslands to arid brush country.

Behavior: A very short, explosive breeding period follows heavy rains, which can occur almost anytime during the year. Hurter's Spadefoots sometimes hybridize with Couch's Spadefoots. In Texas, where habitat differences that normally keep these two species separated are altered by human activities, such mating mistakes are likely. For example, Hurter's Spadefoots may be concealed in sand used to fill the sand traps of golf courses, which are located in the loamier, darker soil preferred by Couch's Spadefoots.

Voice: The advertisement call sounds basically identical to that of the Eastern Spadefoot: a nasal down-slurred *errrrrah!* repeated every five seconds or so. The male calls while floating, arching his body backward and throwing his head upward with each call.

Couch's Spadefoot

Scaphiopus couchii (2 1/4"–3 1/2")

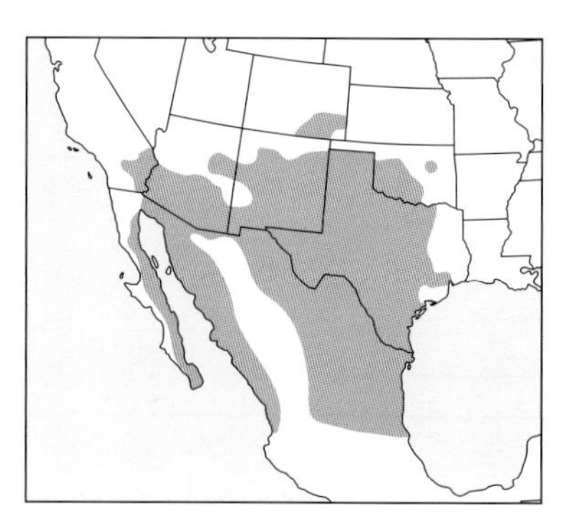

The largest and one of the most colorful of the spadefoots, Couch's Spadefoot is a common southwestern species. It is named in honor of Darius Nash Couch (1822–1897), a Civil War general who had a fascination with natural history and collected the first specimens of this species for the Smithsonian Institution.

Appearance: The most typical color pattern consists of greenish or yellowish netlike markings on a dark brown or black ground color. Some individuals may have a greenish ground color with paler green and greenish yellow markings. Males are typically more evenly colored and duller than females. The spade on the hind foot is elongate and usually curved or sickle shaped (page 254).

Range and Habitat: Found from Texas westward to southeastern California in a wide variety of arid and semiarid habitats ranging from prairie grasslands to mesquite and creosote-bush deserts. Like other spadefoots, Couch's Spadefoot survives dry periods by burrowing into the soil and surrounding itself in a skinlike cocoon to limit water loss during dormancy.

Behavior: Couch's Spadefoots breed after heavy rains in the early winter and spring in the eastern part of the range, and during summer monsoon rains in the West. Breeding may not occur at all during drought years. In the arid Southwest, spadefoots may emerge from their burrows in response to the vibrations of approaching thunder or raindrops hitting the ground. Breeding follows rapidly, and eggs may hatch within just a day of being laid. This species has one of the shortest developmental stages of any of our native frogs and toads; the tadpoles transform about two weeks after hatching. Even so, many tadpoles perish because flooded pools dry up before they can escape.

Voice: The nasal advertisement call is a gagging *errrrrah,* likened by some to the bleat of a sheep or goat. The call is regularly repeated at a rate of about once every five seconds: *errrrrah . . . errrrrah . . . errrrrah . . . errrrrah.*

Plains Spadefoot

Spea bombifrons (1 1/2"–2 9/16")

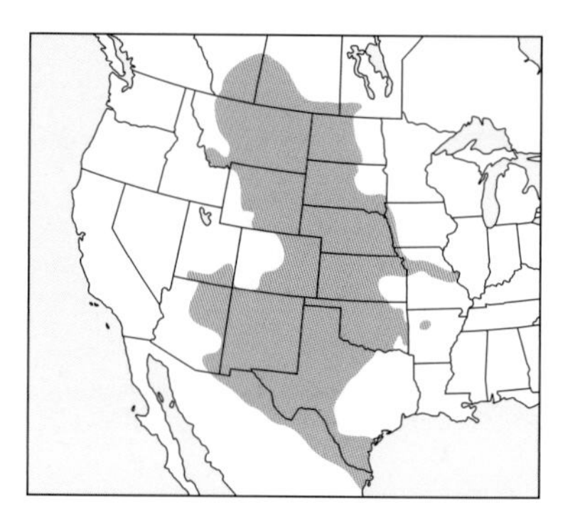

As its common name suggests, the Plains Spadefoot is a resident of prairie grasslands and associated open country. Like other members of its group, it is a fossorial species that is capable of excavating its own burrow but may also inhabit vacant burrows made by small mammals.

Appearance: The typical ground color is some shade of brown, gray, or green, with irregular markings of darker brown or gray. Small wartlike tubercles with yellow or orange tips are scattered on the back and legs. Some individuals have light stripes, two of which may be curved to suggest an hourglass outline, but this is rarely as distinctive as in the Eastern and Hurter's Spadefoots. There is a raised area (boss) between the close-set eyes. The spade is short and wedge shaped (page 254).

Range and Habitat: Inhabits open country such as grasslands, sagebrush, and farmland. Found from Texas to Arizona (including northern Mexico) in the southern part of its range and from Manitoba to Alberta at the northern end. Although they are primarily a plains species, some populations persist in the broad, flat farmlands (formerly grasslands) in the Missouri River Valley as far east as St. Louis. Isolated populations also occur in northwestern Arkansas.

Behavior: Breeding is explosive and occurs after heavy rains in temporary pools from early spring to summer. When not breeding, Plains Spadefoots spend most of their time underground, but during wet spells they may venture aboveground at night to feed.

Voice: The advertisement call is a snorelike bleat or growl repeated once every one to two seconds. Some populations in Arizona, New Mexico, and farther south have very short calls that are repeated rapidly and sound like the yap of a small dog or the quack of a duck.

Great Basin Spadefoot

Spea intermontana (1½″–2½″)

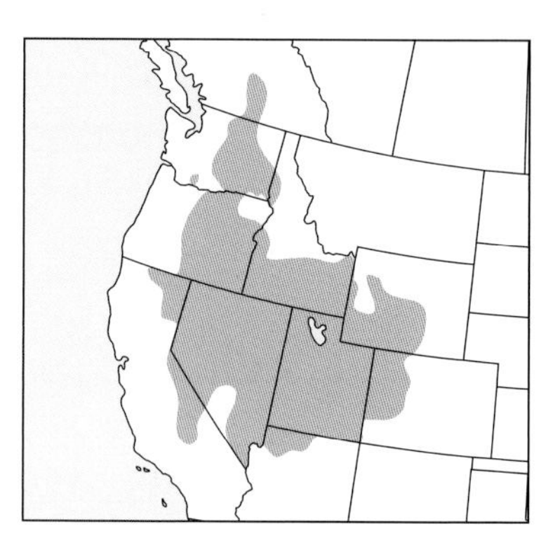

The Great Basin Spadefoot inhabits the arid, high desert lands of Nevada and surrounding states and is the only spadefoot whose range extends into the Pacific Northwest. Because they live where water is scarce, in certain locations this species has benefited from the creation of small artificial impoundments and cattle tanks, which they use as breeding sites in areas that otherwise would remain completely dry.

Appearance: The Great Basin Spadefoot has a bump or boss between the eyes, but it is soft rather than bony as in other species of spadefoots. The spade on the hind feet is wedge shaped. An hourglass-like pattern of light lines is typically present on the back; this pattern contrasts with a darker ground color of gray, dark brown, or olive.

Range and Habitat: Found throughout the Great Basin, including parts of California, Arizona, Colorado, and Wyoming, and extending north into eastern Washington and south-central British Columbia. It occurs not only in sagebrush flats and semidesert habitats but also at higher elevations in open pine forests and moist woodlands, including spruce-fir forests.

Behavior: Most breeding is stimulated by warm spring or summer rains, but some populations are atypical (for spadefoots) in that they may breed in the absence of rain in permanent bodies of water, including clear streams and springs. Calls are given at night and can be heard from a long distance. Typically nocturnal in behavior, Great Basin Spadefoots take shelter in small mammal burrows as well as self-constructed burrows in damp and muddy areas.

Voice: The advertisement call is a hoarse, raspy bleat or snore that is repeated over and over: *waaa . . . waaa . . . waaa . . . waaa . . . waaa.* It sounds similar to the call of the Plains Spadefoot but is often briefer and less snorelike (sounds like a bleat when temperatures are warm and a more drawn out snore when temperatures are cold).

New Mexico Spadefoot

Spea multiplicata (1½"–2½")

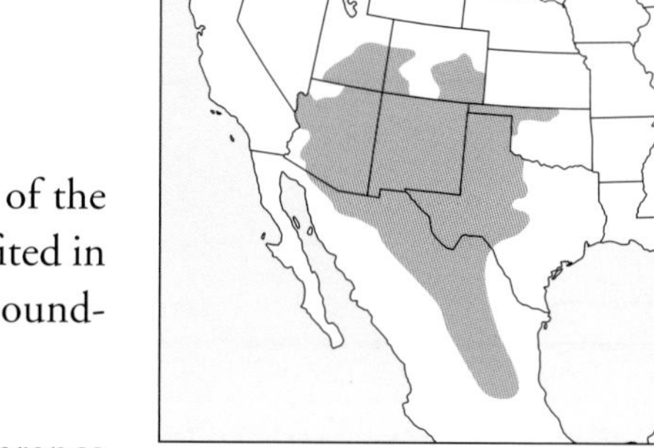

Like other western spadefoots, the New Mexico Spadefoot spends most of the year buried in the soil but emerges to breed after heavy rains. It has benefited in some dry areas from the creation of cattle tanks and other man-made impoundments that provide standing water for breeding.

Appearance: Although it overlaps in range with and is similar in appearance to the Plains Spadefoot (to the east) and the Great Basin Spadefoot (to the north), the New Mexico Spadefoot is easily distinguished by its lack of a boss (bump) between its eyes. The irises of its eyes, which are relatively closely set, are copper colored. The back has a scattering of darker spots or patches that contrast with a gray or dark brown ground color that sometimes has a greenish tinge. Lighter lines sometimes occur on the back, but they do not usually form a distinct hourglass-like pattern.

Range and Habitat: The New Mexico Spadefoot is a southwestern species ranging from Texas to Arizona and northward into Utah and Colorado. It is found in a wide variety of arid and semi-arid habitats, including grassland, creosote-bush desert, and sagebrush desert. It also frequents moister habitats such as open oak and pine forests.

Behavior: Breeding occurs in temporary pools of all kinds after heavy summer monsoon rains, sometimes as late as August. Hybridization with Plains Spadefoots is common. Surprisingly, tadpoles from matings of female Plains Spadefoots with male New Mexico Spadefoots actually develop more quickly than do tadpoles from matings of female and male New Mexico Spadefoots, which puts the hybrid offspring at an advantage in shallow pools that dry quickly. However, the adult survivorship and fertility of these hybrids are likely to be reduced.

Voice: The advertisement call is a metallic, vibrating snore lasting about a second and repeated every few seconds: *waaa . . . waaa . . . waaa.* It sounds very similar to the snore of other western spadefoots.

Western Spadefoot

Spea hammondii (1½"–2½")

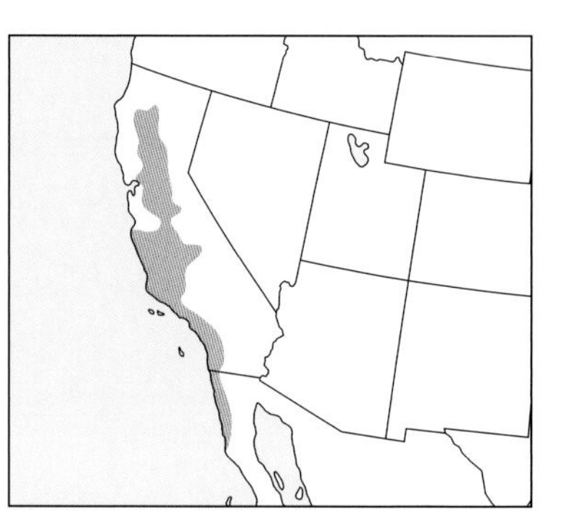

The Western Spadefoot is restricted in range to California and Baja California. It is easy to identify in the field because no other species of spadefoot are found where it occurs. Often occupying areas favorable to human development, it has lost substantial habitat over the last century, and its populations have been reduced to a fraction of their previous numbers.

Appearance: Similar in appearance to the New Mexico Spadefoot, the Western Spadefoot lacks a boss (bump) between the eyes and its eyes are relatively close together. It differs from the New Mexico Spadefoot in having a more elongated spade and pale gold irises (as opposed to copper). The dorsal ground color varies from gray or brown to pinkish brown or light green, and there is usually a scattering of irregular dark spots or blotches. A vague hourglass-like pattern may be present on the back.

Range and Habitat: Found across the lowland central valley area of California, except where bedrock under thin soils limits burrowing. Its range also extends southward into northwestern Baja California. It occupies habitats that are less arid than those used by other species of spadefoots: lowlands, river floodplains, intermittent streams, foothills, and mountains. Within these areas it is typically found where vernal pools regularly occur and where vegetation is sparse.

Behavior: Breeding occurs in winter and early spring in response to rainfall. Unlike typical desert species of spadefoots, it may breed in intermittent streams as well as temporary pools. This species has been the focus of much research on the physiological mechanisms that allow spadefoot tadpoles to develop rapidly. It has been found that when water levels in breeding pools drop, specific cortical hormones from the brain are released; this stimulates secretions by the thyroid and renal hormone systems, which in turn accelerates metamorphosis.

Voice: The advertisement call is a hoarse, nasal snore that is repeated every few seconds: *waaa . . . waaa . . . waaa.* It sounds like the purring of a cat and is similar to the calls of other western species of spadefoots. Choruses are loud and from a distance may sound like someone trimming wood with a handsaw.

Microhylid Frogs and Toads — Family Microhylidae

The family Microhylidae is a diverse group of arboreal, terrestrial, and burrowing frogs and toads represented worldwide by 430 species and 70 genera. In North America, there are only two genera and three species: the Eastern and Great Plains Narrowmouth Toads (genus *Gastrophryne*), and the Sheep Frog (genus *Hypopachus*). These are small, nocturnal, burrowing anurans with triangular, pointed heads and narrow mouths—adaptations for eating ants and termites. Their bodies are rotund, their legs are short, and they tend to run or hop erratically when disturbed. There is a fold of skin across the back of the head. Although moist-skinned and lacking warts, the narrowmouth toads are referred to as toads because of their chunky toadlike appearance.

All three narrowmouth toads are secretive and rarely seen outside the breeding season. They prefer sandy soils in which they can easily burrow. Mating occurs in semipermanent and temporary bodies of water formed after heavy rains. Calling males often hide at the base of dense clumps of emergent grass, making them difficult to find. Their nasal advertisement calls are buzzy whines, reminiscent of the *baa* of a sheep or goat.

Sheep Frog

Eastern Narrowmouth Toad

Gastrophryne carolinensis (7/8″–1 1/2″)

Easily recognized by its small size, rotund body, fat legs, and pointed head, the rather odd-looking Eastern Narrowmouth Toad is the only member of its family that ranges into the Southeast. Highly terrestrial and secretive in habits, it may be found hiding under boards, logs, and other debris in moist areas.

Appearance: In areas of overlap along the western edge of its range, the Eastern Narrowmouth Toad can be distinguished from Great Plains Narrowmouth Toads and Sheep Frogs by its darkly mottled ventral surface. The ground color varies from gray to brown or reddish brown, and individuals can change color. There is often a broad dark area (with scattered black markings) running down the center of the back and flanked on each side by lighter bands. The lower sides, including the sides of the face, are often darkly colored and speckled with white.

Range and Habitat: Primarily a southeastern and lower midwestern species, found from southern Maryland to the Florida Keys and west to eastern Texas, Oklahoma, and southern Missouri (an isolated population occurs in extreme southeastern Nebraska). Found in a wide variety of habitats, including swamps, bottomland forest, pinewoods, prairies, and even coastal dune scrub.

Behavior: Eastern Narrowmouth Toads typically breed after heavy rains occurring anytime from March to October (restricted to spring and summer in northern areas). They prefer temporary pools and other flooded areas but may also breed in permanent water if there are mats of floating vegetation. Males usually hide in grass and under debris when calling and can be very difficult to find, especially when only their snouts protrude from the waterline. They may also call while floating, anchoring themselves to grass and floating vegetation with their front legs and feet.

Voice: The advertisement call is a buzzy, nasal *waaaaaaaaa!* lasting from one to several seconds and sounding like the bleat of a lamb. A soft squeak or peep may precede each call. There is also a distinctive aggressive call—a harsh, throaty *brrrrrrrr,* repeated slowly. People who hear it often think it is the call of a different species.

Great Plains Narrowmouth Toad

Gastrophryne olivacea (7/8"– 1 5/8")

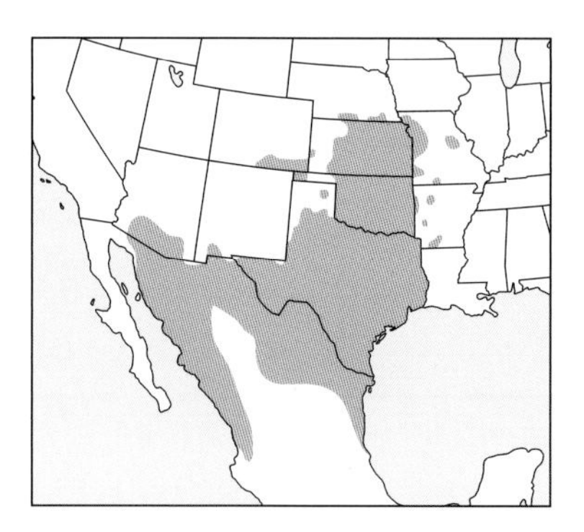

Great Plains Narrowmouth Toads are secretive and difficult to find except when they are breeding. Although these toads do frequent grasslands in the central prairie region, the name *Great Plains* is something of a misnomer because populations also occur in pine-oak woodlands in southern Arizona, desert scrub in New Mexico, and river floodplains in the upper Midwest.

Appearance: Sharing the unique body shape of other members of its family, the Great Plains Narrowmouth Toad can be identified by its light, unmarked belly. Although the young have varied dorsal patterns, most adults are plain-colored gray, tan, or olive, sometimes with a scattering of black spots. Males have dark or light yellow throats, and tubercles on the lower jaw and chest.

Range and Habitat: Ranges from Texas north to Kansas, Missouri, and southern Nebraska. There is an isolated population in southeastern Colorado and the adjacent Oklahoma Panhandle, and northward extensions from Mexico are found in southwestern New Mexico and south-central Arizona. Terrestrial and nocturnal in habits, individuals hide under tree bark and ground debris in moist habitats and seek refuge in the abandoned burrows of rodents and reptiles. In the Southwest, they are known to co-inhabit burrows with tarantulas.

Behavior: Breeding is triggered by rainfall in spring and early summer, but in the arid Southwest it coincides with the arrival of the summer monsoon rains. Breeding occurs in a variety of temporary and permanent bodies of water, including ditches, ponds, springs, and small streams. A skin gland in the male produces a secretion that may help him adhere to the female during mating, and both sexes produce a distasteful, toxic secretion that probably helps protect them from predators. (If you handle this frog, take care to wash your hands before touching your eyes.)

Voice: The advertisement call is a nasal, buzzy bleat lasting one to four seconds; it is preceded by a loud, high-pitched squeak or peep: *peep-waaaaaaaa!* It is easily confused with the call of the Eastern Narrowmouth Toad, but typically has a much louder introductory peep.

Sheep Frog

Hypopachus variolosus (1″– 1 3/4″)

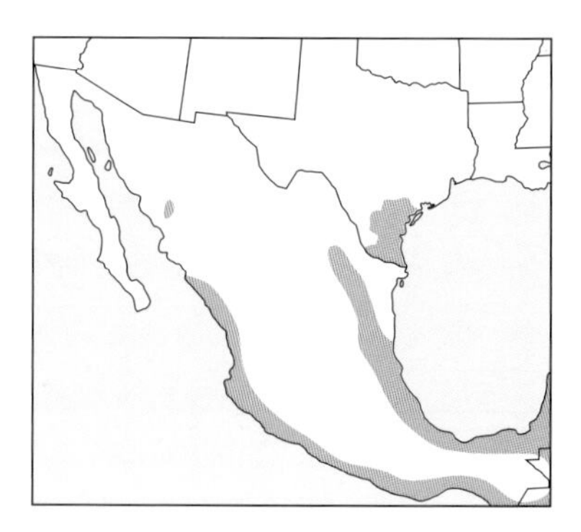

The Sheep Frog gets its common name from its breeding call, which sounds like the bleat of a sheep or lamb. Ranging as far south as Central America, it reaches its northern limit in southern Texas, where it is classified as a threatened species because of habitat destruction. The Sheep Frog has two prominent spades on each heel, which are used for burrowing.

Appearance: Sheep Frogs belong to the same family as our narrowmouth toads, and they share their distinguishing characteristics: rotund body shape, pointed snout, a fold of skin across the back of the head, and fingers and toes without webbing. What makes the Sheep Frog distinctive is a prominent yellow line down the center of the back. There is also another light yellowish line on the darkly mottled undersides. It extends from the tip of the jaw to the lower belly and has diagonal extensions that radiate from the center of the chest to under the arms.

Range and Habitat: Abundant over much of Central America and Mexico. Found in the United States from the lower Rio Grande Valley north to the Corpus Christi area (it seems to be expanding its range northward). Frequents areas where the humidity is usually high, and in the United States does not range more than about fifty miles inland from the Gulf of Mexico. Habitats include open woodlands, pasturelands with short grass cover, tropical thorn scrub, and savannas. Individuals occupy burrows by day and may also take refuge under logs and other debris.

Behavior: In southern Texas, breeding is triggered by heavy rains occurring during relatively warm periods from March to September. Sheep Frogs use a variety of breeding sites, temporary to permanent. The male often calls from a floating position, sometimes clasping a grass or bush stem with his front feet. Sheep Frogs are ant and termite specialists (as are all narrowmouth toads), but they also eat flies.

Voice: The advertisement call is a nasal bleat lasting nearly two seconds and repeated at variable intervals, sometimes with many seconds between calls. It sounds similar to the nasal call of the Great Plains Narrowmouth Toad, with which it overlaps in range, but is more robust and considerably lower in pitch.

Neotropical Frogs — Family Leptodactylidae

A small number of neotropical frogs are found in certain areas in the southern United States, especially in Texas and Florida. There are seven species altogether, including three chirping frogs, the Barking Frog, the Greenhouse Frog (introduced), the Puerto Rican Coqui (introduced), and the White-lipped Frog. All but the last species lay their eggs on land, with the eggs hatching into miniature terrestrial adults. The White-lipped Frog, which has aquatic tadpoles, is unique in that it uses its hind legs to whip secretions into a foam nest into which the eggs are laid.

In this book, we follow the long-standing tradition of placing these neotropical frogs in the family Leptodactylidae, a grab bag of sorts for a diverse set of more than 1,100 species that are found mostly in South America, Central America, Mexico, and the West Indies. Recent studies, however, suggest that nearly all our species, with the single exception of the White-lipped Frog, should be placed in the family Brachycephalidae. One new generic name change that we have adopted is placing the Barking Frog in the genus *Craugastor,* rather than leaving it in the genus *Eleutherodactylus* where it was formerly placed. We include our native chirping frogs in the genus *Eleutherodactylus,* while some consider them to be members of the genus *Syrrhophus.* Given that changes in genera and family designations occur with some regularity in the field of taxonomy, we are grateful that the common names of our frogs and toads remain relatively stable.

Cliff Chirping Frog

Barking Frog

Greenhouse Frog

Eleutherodactylus planirostris (5/8"– 1 1/4")

A West Indian species abundant in Cuba, the diminutive Greenhouse Frog has been introduced into Florida and other locations in the Southeast as well as into Hawaii, where it is now well established. In Florida, it is common around greenhouses, gardens, and nurseries, where individuals may be found hiding under boards, mulch, stones, and flowerpots.

Appearance: There are two main color phases, one having prominent light longitudinal stripes and the other irregular markings on the back and sides. There is usually a light orange or tan band connecting the eyes. The ground color varies from light to dark brown or reddish brown, and the skin has numerous bumps or tubercles.

Range and Habitat: Thought to have been introduced in the late 1800s and already abundant in central Florida by the 1940s, the Greenhouse Frog has rapidly expanded its range. Now found throughout peninsular Florida with isolated pockets occurring in Georgia, Alabama, the Florida Panhandle, and Louisiana. More recently, it has established itself on two Hawaiian islands, Oahu and Hawaii, where it is considered an invasive species. In Florida, Greenhouse Frogs are found in a wide variety of habitats, from urban greenhouses to remote hammocks and other natural areas.

Behavior: The Greenhouse Frog is a terrestrial breeder and can be heard calling mostly from April to October. Eggs are laid in moist debris on the ground, and the larval stage is passed within the egg. Individuals take refuge by burrowing into damp leaf litter. The Greenhouse Frog is also a frequent inhabitant of cave entrances and gopher tortoise burrows.

Voice: The advertisement call is a liquid, metallic chirp or *p-dit* that could easily be mistaken for the tinkling chirp of a cricket. Calls both day and night, especially during hot rainy periods and when large storm systems are passing.

Puerto Rican Coqui

Eleutherodactylus coqui (1″–2¼″)

Widespread in moist lowlands and mountains of Puerto Rico, the Puerto Rican Coqui is a small to medium-sized tropical frog that has been introduced to southern Florida, Hawaii, and the Virgin Islands. It probably arrived at these locations on imported ornamental or nursery plants, most likely as egg masses hidden among leaves. Coquis are named for their vibrant and musical breeding call, which sounds like *co-kee!*

Appearance: The ground color is usually light brown or tan, and there may be a scattering of dark spots. Individuals sometimes have prominent dorsolateral stripes and light bands between the eyes. The toes lack webbing but have large discs on their undersides that help in climbing.

Range and Habitat: In the continental United States, found in and around greenhouses in southern Dade County, which is in extreme southern Florida. Coquis first appeared in the late 1990s in Hawaii (which doesn't have any native frogs), and they are now resident on all the islands. The tourist industry there has become worried that the loudness of their calls will keep vacationers away.

Behavior: Coquis can breed year-round, but most activity in Florida occurs during warm and humid periods from late spring until early autumn. As with other members of its genus, the coqui is a terrestrial breeder. Males call from the ground and from elevated perches. Females lay eggs in protected areas among leaves, and males guard them.

Voice: The advertisement call is a melodic, whistled *co-kee!*, with accent on the second, higher-pitched syllable. The introductory *co* probably functions as a threat to neighboring males, while the *kee!* attracts females. During vocal encounters, males often drop the second part and threaten one another with *co* calls.

Cliff Chirping Frog

Eleutherodactylus marnockii (3/4″– 1 1/2″)

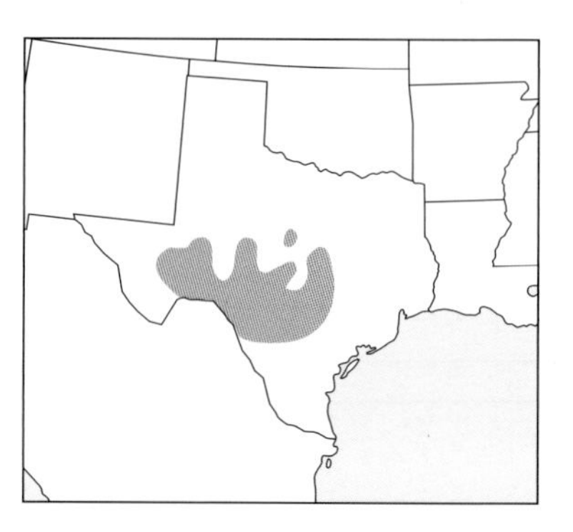

One of three species of chirping frogs found in Texas, the Cliff Chirping Frog inhabits rocky forested areas throughout much of the south-central part of the state. It is a colorful little frog with a large head and eyes in proportion to its body size and might easily be mistaken for a juvenile. Its flattened body shape allows it to slip into narrow crevices between rocks.

Appearance: The dorsal color is a vibrant yellow-green or yellow-brown, with contrasting darker spots or blotches. The side of the face, from snout to tympanum, is brownish and lacks any spotting. The underside is light with a purplish, reddish, or brownish tint. Tiny squarish toe pads aid in climbing.

Range and Habitat: A denizen of juniper-oak forest, found from the Edwards Plateau west to the Stockton Plateau. Always found in association with rocks and boulders, especially limestone, where individuals take shelter in cavities, crevices, and caves. Also frequents suburban areas and city parks, where it may be found hiding under boards, bricks, and other debris.

Behavior: The Cliff Chirping Frog is terrestrial and nocturnal in habits. On rainy nights, it is often seen rapidly scurrying across the road, like a small mammal (chirping frogs can run as well as leap and hop). Most breeding takes place during rainy periods from April to May, but a second breeding season may occur from September to October. Males call both day and night during peak periods. Females dig shallow trenches in which they lay their eggs, and one or the other parent may guard the eggs until hatching.

Voice: Males have two types of calls. The more common call is a brief, musical chirp that is regularly repeated every three to five seconds. The other call is a short rattling trill. Typically, a male will chirp leisurely for a number of minutes and then suddenly give an excited series of about five chirps, followed by five or more of the trills.

Spotted Chirping Frog

Eleutherodactylus guttilatus (3/4″– 1 1/4″)

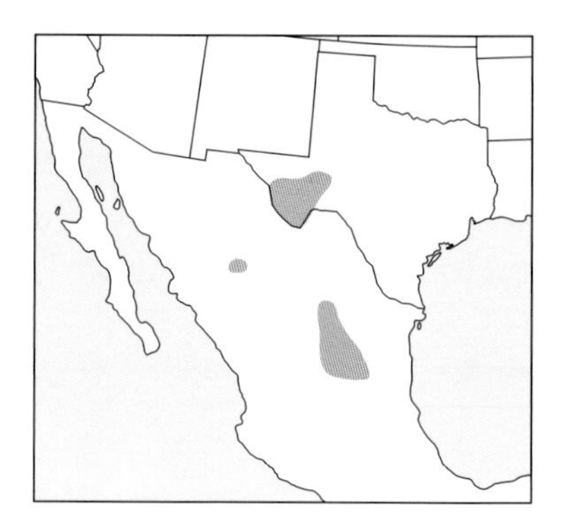

Extremely secretive and difficult to find, the Spotted Chirping Frog is restricted to the Big Bend region of Texas, where it is locally common to abundant. It looks much like the Cliff Chirping Frog, but the ranges of the two species do not overlap, except for a possible contact zone to the northeast of Big Bend. No detailed genetic studies are available, and Spotted and Cliff Chirping Frogs might eventually be considered the same species.

Appearance: Ground color ranges from yellow to olive, tan, or brown, and there is a mottling of dark spots or blotches (in some individuals, dark areas predominate and are mottled with yellowish markings). The sides and belly range from light to dark. The body shape is somewhat flattened (like other chirping frogs'), and there are tiny squarish toe pads and no webbing on the feet.

Range and Habitat: Found in the Big Bend region of Texas and south into Mexico. Inhabits rocky outcrops, cliffs, hillsides, caves, mines, limestone bluffs (along the Rio Grande), and even man-made rock retaining walls — wherever there are cracks and crevices in which to hide.

Behavior: The Spotted Chirping Frog is a terrestrial breeder. Breeding activity occurs mainly with the onset of summer rains in June and July, when males can be heard calling from cracks and crevices among rocks from ground level up to about five feet. They are active primarily at night. Little more is known concerning their breeding behavior.

Voice: The calls are nearly identical to those of the Cliff Chirping Frog. The common call is a brief musical chirp that is regularly repeated every few seconds. Another call is a short rattling trill.

Rio Grande Chirping Frog

Eleutherodactylus cystignathoides (5/8″– 1″)

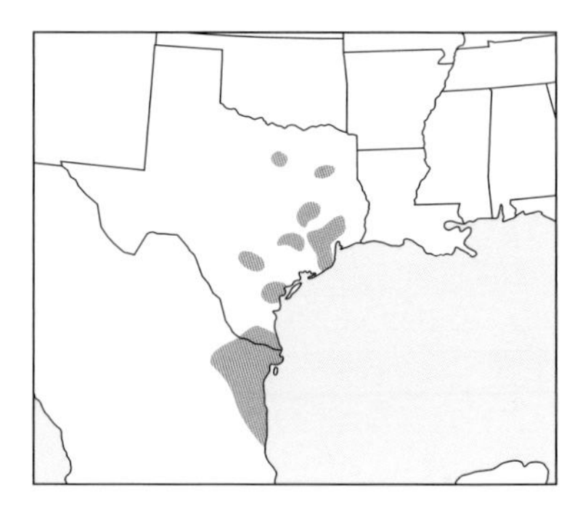

The Rio Grande Chirping Frog is a Mexican species that occurs naturally in the United States only in extreme southeastern Texas. However, it has been inadvertently introduced (in potted plants and nursery stock) to a variety of other locations in eastern Texas, especially various urban areas, where it has become common. In spite of its large numbers, the Rio Grande Chirping Frog is difficult to find—hearing its call during breeding season is one of the best ways to verify its presence.

Appearance: In comparison to the other two species of chirping frogs, the Rio Grande Chirping Frog is rather nondescript. Ground color varies from brown or gray to olive or yellow-green. Spots are usually absent or poorly defined. Like the bodies of other chirping frogs, the Rio Grande Chirping Frog's body is elongate and flattened, and the feet lack webbing and possess tiny toe pads.

Range and Habitat: The natural range is in northeastern Mexico and the lower Rio Grande Valley of Texas. Introduced populations occur in Dallas, Houston, Corpus Christi, San Antonio, Kingsville, Tyler, Huntsville, and La Grange. The preferred habitat is thick vegetation along streams and around pools and ponds. Rio Grande Chirping Frogs also thrive in the presence of humans, especially in areas that are kept artificially moist, such as well-watered lawns and gardens, where they may hide under rocks, stones, boards, and other objects. They are agile and fast, and they escape by running or leaping.

Behavior: Rio Grande Chirping Frogs are terrestrial and nocturnal in habits. Breeding occurs mostly in April and May in south Texas, and from April to July in northern areas (such as around Houston). Males call from the ground and from low vegetation, mostly at night but also during the day, especially after rains and during humid spells. Eggs are laid in the soil and in other moist, protected areas and are possibly tended until hatching by one of the parents.

Voice: The advertisement call is a high-pitched, insectlike (or birdlike) *peet,* often given in groups of two to four: *peet-peet-peet-peet.* Males also produce soft trilled notes. Calls of this species are higher in pitch than the calls of the Cliff Chirping Frog.

Barking Frog

Craugastor augusti (2″– 3 3/4″)

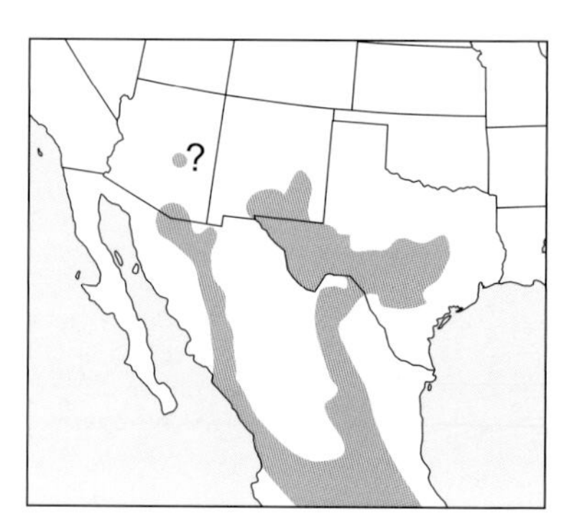

The Barking Frog is named for its vocalization, which from a distance may sound like the bark or yelp of a dog. Like other tropical frogs in its group, it is a terrestrial breeder, and it lays its eggs on land. Barking Frogs are secretive and difficult to find, and little is known about their natural history.

Appearance: A Barking Frog is readily identified by a prominent inverted **U**-shaped fold of skin across the back of the head that extends a short distance along the sides. A similar fold of skin called the ventral disc is found on the belly. The ground color may be gray, gray-brown, olive-brown, reddish brown, or tan, and there are darker blotches on the back and some banding on the legs. Prominent bumps or tubercles occur on the feet, and the toes lack webbing.

Range and Habitat: The Barking Frog ranges from southern Mexico north into the United States, where it is found patchily from central Texas to southeastern New Mexico. Populations also occur in extreme southern Arizona and there is an old, unconfirmed record from central Arizona. In Texas and New Mexico, Barking Frogs frequent the rocky slopes and outcrops of canyons, especially limestone bluffs along creeks, and are also found in creosote-bush flats. In southeastern Arizona, they occur on rocky slopes of mountains at altitudes of about 4,200–6,200 feet. Wherever they are found, Barking Frogs find shelter under rocks and boulders and also in crevices, caves, and rodent burrows.

Behavior: The breeding season in Texas is in March and April. In Arizona, breeding occurs more explosively, with the arrival of the summer monsoon rains in late June and July. Breeding takes place on land, probably under boulders, in small caves, and in other moist, protected areas. Females are thought to stay with the eggs until they hatch.

Voice: The advertisement call is often described as sounding like the distant yelp of a dog, but when heard at close distance it is actually a harsh, guttural *warrrr* (with rolled *r*'s) repeated every two seconds or so. When in close proximity to one another, interacting males closely alternate their calls.

White-lipped Frog

Leptodactylus fragilis (1 3/8"–2")

Named for the white line on its upper lip, the secretive and nocturnal White-lipped Frog of Mexico is our only native frog that builds a special foam nest for its eggs. In the United States, it is restricted in range to the lower Rio Grande Valley of Texas.

Appearance: The White-lipped Frog is streamlined in appearance and might easily be mistaken for a small ranid (true frog) were it not for a prominent **U**-shaped ventral disc. The ground color may be brown, gray, tan, or olive, and there are some black markings on the back and sides. Dorsolateral folds are present but sometimes indistinct, and the legs are usually barred.

Range and Habitat: Ranges from Venezuela northward through Central America and Mexico to a few locations in the lower Rio Grande Valley of Texas. Considered threatened by the State of Texas, where there is concern that use of organophosphate fertilizers could lead to its extirpation. Favors grassy areas next to semipermanent lakes, ponds, and pools, but also found in and around ditches and canals in agricultural areas.

Behavior: Breeding occurs during rainy periods, usually in late spring and summer. It may not breed at all during dry years. Males call from depressions and shallow burrows under grass clumps, dirt clods, and debris. The male excavates a special brood chamber, and eggs are laid in a foam nest made of frothy body secretions, which helps protect the eggs from drying. After hatching, the larvae remain sheltered in the liquid center of the nest but escape into surrounding water after heavy rains.

Voice: The advertisement call is an up-slurred whistled *prrIT* that is accented at the end. Calls are delivered in a brisk series at a rate of about two times per second: *purrIT-purrIT-purrIT-purrIT-purrIT.*

Tailed Frogs — Family Ascaphidae

The tailed frogs of the genus *Ascaphus* are considered to be living representatives of the most ancestral of all living frogs. Formerly thought to be a single species, two tailed frogs are now recognized, and both inhabit fast-flowing mountain streams in the Pacific Northwest. Their closest relatives are primitive frogs of the genus *Leiopelma* from New Zealand — the only frogs found in that country. Both genera share the ancestral traits of nine vertebrae in front of the sacrum and special tail-wagging muscles at the rear end. In this book, we include the tailed frogs in the family Ascaphidae, of which they are the only two members. But note that recent studies present a good case for placing them, along with their three New Zealand relatives, in the family Leiopelmatidae.

Tailed frogs are named for the male's "tail," which functions as a copulatory organ for internal fertilization of eggs (internal fertilization occurs in a few other frogs, but none possesses a copulatory organ). The male's tail is derived from his cloaca and includes muscles and skeletal tissue. Amplexus is inguinal, meaning that the male grasps the female around her waist. It is considered a primitive behavior, and in order to insert his tail, the male must repeatedly assume an awkward position — a posture that can be maintained for a short period of time only and during which a limited number of eggs can be fertilized.

Close-up of "tail" of male Rocky Mountain Tailed Frog

Female Pacific Tailed Frog

Pacific Tailed Frog (male)

Rocky Mountain Tailed Frog and Pacific Tailed Frog

Ascaphus montanus (1″–2″) and Ascaphus truei (1″–2″)

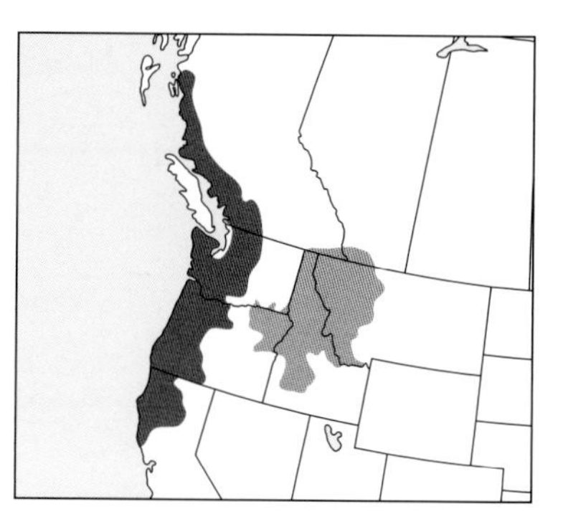

The prominent "tail" of a male tailed frog is really a sex organ used to fertilize eggs internally, a feature unique among frogs and toads and an adaptation to breeding in fast-flowing water. Tailed frogs do not vocalize and lack external eardrums, traits that probably evolved because loud stream noise prevented them from communicating with sound. There are two species that look alike but are genetically distinct and do not overlap in range: the Rocky Mountain Tailed Frog *(A. montanus)* and the Pacific Tailed Frog *(A. truei).*

Appearance: Flat-bodied and squat, tailed frogs are unusual in having ribs (which most frogs lack) and more vertebrae than other frogs. Their color may be olive, gray, brown, or reddish brown, and there is often a greenish or yellowish triangle between the eyes and the snout. There is a dark stripe running through each eye, and the eyes have vertical pupils. The tadpole has a wide head and a large ventral sucker that is used to hold on to rocks in fast-moving water.

Range and Habitat: Rocky Mountain Tailed Frogs (orange on map) occur in Alberta, Idaho, Montana, and extreme eastern Washington and Oregon. Pacific Tailed Frogs (red on map) occur in the Pacific Northwest in western British Columbia, Washington, Oregon, and northern California. Both species are denizens of cold, fast-flowing streams, especially in old-growth montane forests.

Behavior: Tailed frogs are mostly nocturnal. They hide under rocks and logs in and near streams during the day. Mating takes place in early autumn, when males develop nuptial pads (padlike structures where the thumb and palm meet) and black wartlike tubercles along their sides, arms, chests, and chins that help them grasp females during mating. After fertilization, the sperm is stored in the female's oviduct and used to fertilize eggs when they are laid the following summer. Tadpoles take two to five years to transform, depending on elevation and latitude, and juveniles can take up to seven or eight years to become sexually mature.

Rocky Mountain Tailed Frog

Voice: Neither species is known to vocalize, and the exact mechanism by which the sexes find each other during mating season is unknown.

Pacific Tailed Frog

Burrowing Toads—Family Rhinophrynidae
Tongueless Frogs—Family Pipidae

BURROWING TOADS

The Mexican Burrowing Toad *(Rhinophrynus dorsalis)* is the only living member of the family Rhinophrynidae. Ranging into the United States in extreme southern Texas only, it is a bizarre-looking species with an egg-shaped body, short legs, and spadelike hind feet. As its common name suggests, the Mexican Burrowing Toad stays underground for most of its life, coming to the surface only after heavy rains in order to feed and breed. Despite the species' fossorial habits, the Mexican Burrowing Toad's closest relatives are members of the family Pipidae (see below), which are highly aquatic.

Mexican Burrowing Toad

TONGUELESS FROGS

The African Clawed Frog *(Xenopus laevis)* is an introduced species that has become established in southern California. It is a member of the family Pipidae, the tongueless frogs, which includes five genera and thirty species found in tropical South America and Africa. Pipids are a very ancient group whose fossil record extends back to the Lower Cretaceous Period, 120 million years ago. Some species are rather grotesque in appearance, with flattened bodies and puffy, undulated skin. Pipids lack tongues and are strictly aquatic as adults. Many live in murky water, where they use their well-developed lateral-line organs to detect subtle movements of nearby prey. The calls of a number of species, including the African Clawed Frog, have a mechanical clicking or popping quality and are produced by a mechanism unique among frogs and toads—the frog activates the larynx directly by contracting muscles rather than using air from the lungs to excite the vocal cords.

African Clawed Frog

Mexican Burrowing Toad

Rhinophrynus dorsalis (2″–3½″)

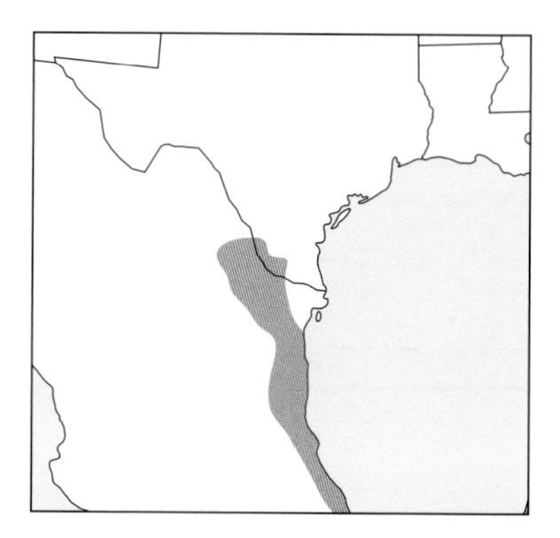

One of the oddest-looking of all our frogs, the Mexican Burrowing Toad inflates like a balloon when calling or when alarmed. Its generic name, *Rhinophrynus,* translates to "snout toad," a reference to its blunt and rounded snout. This species is unusual among our native frogs in having a tongue that is anchored at the back, as in mammals—all our other native frogs have a front-anchored tongue that is flipped forward when extended.

Appearance: The body is rotund and the head is triangular, much like the shape of narrowmouth toads. Burrowing is facilitated by horny spades on each heel, plus short stumpy forelimbs and legs. The ground color is black, and there is a prominent orange or reddish stripe down the center of the back. There is also a scattering of tiny white-tipped turbercles across the back and legs, and there may be large white to orange blotches on the sides.

Range and Habitat: Widespread in Mexico and northern Central America but local and rare in the United States, where small numbers are found at a few locations in the lower Rio Grande Valley of Texas, at the extreme northern limit of the range (the species was first discovered in the United States in 1964). Favors arid areas with loose soils, including cultivated fields.

Behavior: Little is known about the natural history of the Mexican Burrowing Toad. Most of its feeding probably occurs underground in ant and termite nests, but it may venture from its burrow to forage on humid nights. Breeding can occur at any time during the year, but only after torrential rains from tropical storms and hurricanes, when males congregate in flooded pools. Most calling occurs in and next to breeding pools, although males sometimes call from their burrows. When a floating male sounds off, the sudden inflation of his internal vocal sacs causes his head to pitch upward with each call.

Voice: The advertisement call is a nasal, moaning *whooooahhhh* with a rising inflection; it lasts about one and a half seconds. Males usually give long series of calls with several seconds of silence between each call.

African Clawed Frog

Xenopus laevis (2″– 5 5/8″)

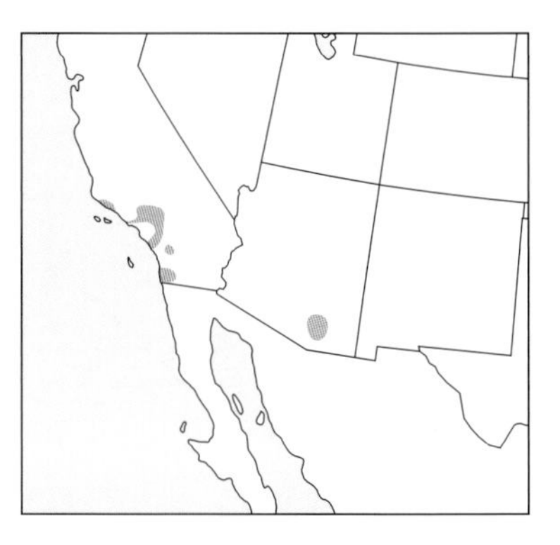

Native to South Africa, the African Clawed Frog is an introduced and invasive aquatic species that was used in the 1940s and 1950s for human pregnancy testing. They were also sold as pets. Named for the claws on their hind toes, clawed frogs either escaped from captivity or were released in the wild in California and other scattered locations. They are omnivorous and can be voracious predators, often using their claws to hold or rip apart prey.

Appearance: Resembling a very large narrowmouth toad, the African Clawed Frog has a flattened body with a small head and a blunt snout. The forefeet lack webs, and the hind feet are fully webbed with sharp black claws located on the three inner toes. The ground color ranges from gray to olive, brown, or yellow-brown, and there is usually a scattering of darker or lighter spots or blotches on the back.

Range and Habitat: The main areas of permanent establishment are in southern California, where populations have expanded over the last forty years in numerous drainages from Santa Barbara to San Diego. While still sold in pet stores over much of the country, clawed frogs have been banned from the pet trade in a number of states, including California, because of concerns about their impact on native species. They are found in a variety of permanent aquatic habitats, including drainage ditches, man-made ponds and other impoundments, flood-control channels, sewage lagoons, and stagnant pools.

Behavior: Most breeding occurs from March through June. Males call while underwater to attract females. Males lack vocal cords and vocal sacs but are able to produce sounds by using muscles in their larynxes to suddenly snap apart the cartilage. Females respond to the males' calls with their own sounds that indicate their readiness to mate. Both sexes lack distinct eardrums (tympana).

Voice: The advertisement call is difficult to describe. While males make occasional ticking notes, they periodically give long mechanical-sounding outbursts that may last ten seconds or more. These outbursts include buzzy scrapes that alternate with metallic rattles.

Finding, Observing, Catching, and Keeping

Finding frogs and toads is a pleasurable pastime and is usually not very difficult to accomplish, as long as you go to the right habitats at the right times of year and have the willingness to search patiently. It's always a good idea to know as much as possible about your quarry. Find out what species are supposed to be in your area. Read up on their natural history. Talk to others who know more about the subject than you do. Because many anurans live in and around water, think about your surroundings and the locations of lakes, ponds, swamps, rivers, and streams. Bring a flashlight for nighttime observation. Close-focusing binoculars will also come in handy, for use both day and night. With a little effort, you will soon find some great spots to look for frogs and toads, and ample opportunities for observation will no doubt come your way.

During the day, a frog or toad may not be doing much—perhaps hopping or leaping and swimming away, or just sitting there breathing. If you're patient, and lucky, you might witness one catching a flying insect or pursuing some other prey. And very rarely, you might observe an interaction with another of its kind. Listen also for loud moans or screams and even subtle squeaks. Follow the sound, and you may discover a frog or toad that has been captured and will soon be swallowed by a snake or other predator.

Throughout the warmer months, searching for anurans at night will usually yield great results. It is often easy to approach individuals sitting along the shorelines of lakes, ponds, or streams. In addition, many species move about on rainy nights and can be found crossing blacktop roads (they cross all roads but are easier to see on shiny pavement). So when it rains, grab your field guide and head to the country. Drive slowly and look for movement as frogs and toads of all sizes hop or walk across the road. Unusually heavy rains will bring out secretive species such as spadefoots and narrowmouth toads, which spend most of their time in hiding. Driving blacktop roads on rainy nights is an excellent way to find out what frogs and toads are in your area.

Undoubtedly the best time to observe frogs and toads is during the breeding season, especially when choruses are loud and breeding is at a peak. On the day before your adventure, review the calls of the species in your area. Then, at nightfall, head for the wetlands. Bring rain gear. Wear hip waders or high boots. Better yet, wear old tennis shoes and prepare to get wet and muddy. Don't forget your flashlight. Be willing to enter into the world of the frog, wet skin and all.

Not only will you be able to locate and observe calling males, you will also find pairs in amplexus and females laying eggs, and you can home in on social interactions by listening for aggressive and release calls. Dense toad choruses offer great opportunities. Find a place to sit at the edge of a breeding pool. Turn off your flashlight and listen. When you hear chirping sounds, shine your light in that direction, and you're likely to see two struggling male toads, one squeaking and irritated because the other has tried to mount it. The denser the chorus, the easier it is to observe interesting social behavior.

Over much of North America, frogs and toads are dormant during the cold months and then become active in late winter or early spring, coinciding with the arrival of the warm rains. Spring Peepers, chorus frogs,

"Blue Frog"— a Green Frog lacking the yellow pigment xanthin

and Wood Frogs may start to call on warm nights before the ice has fully melted from the pools. Soon after, there will be an explosion of activity. Cricket frogs, toads, treefrogs, narrowmouth toads, and various true frogs will chime in, each according to its own seasonal clock. And nearly anytime, a huge downpour might set off a chorus of spadefoots or some other species that requires major rainfall events. From spring to summer, and sometimes beyond, frogs and toads will be breeding somewhere.

When you're visiting a breeding chorus, you may find it difficult to judge your distance from a calling male. A solution is for two people to triangulate the caller's position using flashlights. It is important to approach at right angles to each other. Once you each get a good fix on the caller, point the light in its direction; the caller will be where the two beams cross. In the absence of a partner, you can use sticks to do the pointing. When trying to locate loud and piercing callers, such as Spring Peepers, you might be surprised at how far away the frog is. It may sound like it's only a few feet distant, but it may actually be ten or fifteen feet away. Even when you locate a caller, the male can still be extremely difficult to see. This is especially true of chorus frogs, narrowmouths, Oak Toads, and other species whose males often call from within clumps of emergent grasses.

It is very exciting to watch a singing male as he inflates his vocal pouch and calls. Frogs and toads typically quit calling when lights shine on them, so after you find a caller, turn the light off and then wait for the male to start again. When he does, quickly turn your light back on—often he will continue calling, at least for a few seconds, giving you a good view. Dim light, such as a flashlight covered with a red lens to cut down on the light, may also be effective. You might consider carrying two flashlights, a bright one for locating frogs and toads and a dim one for close observation. The very best conditions for observation occur in dense and nearly deafening choruses, when males are so excited that mild disturbances do not affect them. Use earplugs if necessary, and wade into the middle of the cacophony. Be like a frog, and you'll be surprised at what you find.

CATCHING

Kids, and many adults too, love to catch frogs and toads. It is almost a basic instinct to chase after a leopard frog as it bounds away, to try to snatch a Bullfrog at water's edge before it leaps and splashes, or to scoop up a little toad found hopping in the woods.

So what if we told you that catching frogs and toads is perhaps not such a good idea or, rather, that it's a good idea sometimes and a very bad idea at other times? Well, that happens to be the truth of the situation. Increased awareness and knowledge of our native frogs and toads has revealed that certain species are experiencing severe and alarming declines, while others have peripheral populations that are in danger of extirpation. Disturbing these particular frogs and toads has a negative impact on their populations and speeds the process of their demise.

There are also major legal issues. Many anurans are now protected by state or federal law—not only endangered species, but others as well. Penalties for illegal possession can be severe, and ignorance of the law is not an acceptable excuse. Before taking any frog or toad captive (adults, eggs, or tadpoles), it is prudent to research the legality of doing so. This information can be obtained by contacting your state department of conservation or your local wildlife service.

Legal issues aside, catching common or abundant species can certainly be a harmless pursuit. But it is important to take simple precautions, such as washing your hands before and after, not squeezing too hard, and releasing your captive soon after the catch, at the same place it was found, just like a fisherman gently releasing a fish back into the water.

Washing your hands before and after is a very important rule to follow. When absorbed through an amphibian's porous skin, topical insecticides, perfumes, hand lotions, soaps, and medications can cause nearly instantaneous death of any frog or toad. In addition, frogs and toads produce glandular secretions that may cause human discomfort if ingested or brought in contact with the eyes or mucous membranes. So be certain to wash your hands and rinse them carefully in cold water before and after handling, both for your sake and the sake of your quarry.

The art of catching anurans might come naturally to some, but many will need practice to get good at it. Certainly, having a quick hand is essential, but it is equally important to have a gentle touch. Frogs and toads are easily damaged if they are squeezed too hard. And holding a frog's leg while it frantically tries to escape can lead to serious and irreversible injury. So it is important to be firm yet gentle. Hold large specimens around the waist, not by the legs. You can cup small frogs and toads in your hand or gently hold them between two fingers to get a good view. A fisherman's net may come in handy, especially for aquatic species, allowing you to scoop a frog from the water's bottom as it hides beneath a leaf or buries itself in shallow mud.

KEEPING

So now you've caught a frog or toad. You've experienced success, but what are you going to do with it? We strongly advise that you take a good look at it and then let it go, right where you found it. Why complicate things for yourself and your captive by taking it somewhere else? But what if you want to take it home so that you can show it to others and observe it? While this is rarely in the best interest of the frog or toad and can possibly contribute to the spread of diseases such as chytrid fungus, it will probably do little harm if it's a common or abundant species that you can legally possess and if you're willing to assume full responsibility for its well-being. Yes, many a hapless frog or toad has been hauled home and handled improperly, usually perishing outright within a week or two or else escaping into a house where it soon shrivels up and dies. Should you insist on taking a frog or toad home, here are two rules to remember: (1) keep your pet happy and healthy during its brief sojourn in your home, and (2) release it within a day or two at the place where you found it, in its favored habitat where it will be able to survive.

While your amphibian pet is under your care, keep it moist and in a container with a tight-fitting lid so that it won't escape into your house. Don't put the container in the sun, or your pet will overheat and die. You need not worry about feeding it because nearly all frogs and toads can get along just fine for several weeks without food. But of course it's fun to watch them feed, and during the summer months it is easy to collect flies, earthworms, termites, and the like, so that your pet has a smorgasbord of food items to choose from. While it is not that difficult to take good care of your pet, we nonetheless recommend releasing it within a day or two, preferably in the exact location from which it came.

Related to capturing adults is collecting eggs or tadpoles and raising them until they transform into adults. This can potentially provide a great lesson about metamorphosis and allow kids and adults to observe firsthand an exciting transformation. While we do not provide guidelines for the care of eggs and tadpoles, many books are available that shed light on this topic. One obvious concern is what to do with adults once the tadpoles have transformed. As a general rule, do not release long-term pets or classroom animals back into the wild. Captive frogs might pick up infections from humans or other pets, and the release of diseased individuals into the wild can have devastating effects on natural populations.

In this book, we do not include information about permanently keeping frogs and toads as pets. There are excellent books on the subject, and we especially recommend *Frogs, Toads, and Treefrogs: A Complete Pet Owner's Manual,* by R. D. and Patricia Bartlett. The key to maintaining the health and well-being of your pet lies in proper husbandry, which means providing housing, care, and food that is appropriate to the species being kept. Countless frogs and toads have perished because their well-meaning keepers grew bored and neglectful and failed to provide proper care. Keeping frogs and toads permanently involves a great deal of knowledge and responsibility and an unwavering commitment to their well-being. The Bartletts' book provides excellent advice on these matters.

Pet frogs and toads are best obtained through purchases from pet stores or breeders who maintain colonies of selected species. Breeders, who often specialize in uncommon or rare species, are easily located with the help of the Internet. While many breeders and suppliers are responsible and operate well within the law, there are a lot of illegal activities, both nationally and internationally, in the pet trade. Removing large numbers of frogs and toads from the wild to sell as pets can unquestionably pose a threat to native amphibian populations. Also, animals moving through pet-trade channels may contribute to the spread of chytrid fungus, a serious and often fatal frog disease (see Conservation Issues, page 32). For these reasons, we strongly advise people to buy captive-bred rather than wild-caught anurans and not to buy animals that come from outside the country.

There are many serious pet owners who excel at keeping their pets healthy and who help educate the public concerning the conservation of our native frogs and toads. Teachers in classroom settings can make excellent use of captive frogs and toads, along with tadpoles and eggs, to create educational experiences that have very positive outcomes. In all these cases, however, we advise keeping the well-being of the frogs and toads in the forefront of the mind. Make choices that will lead to a greater understanding of these incredible creatures and that will help protect them for generations to come.

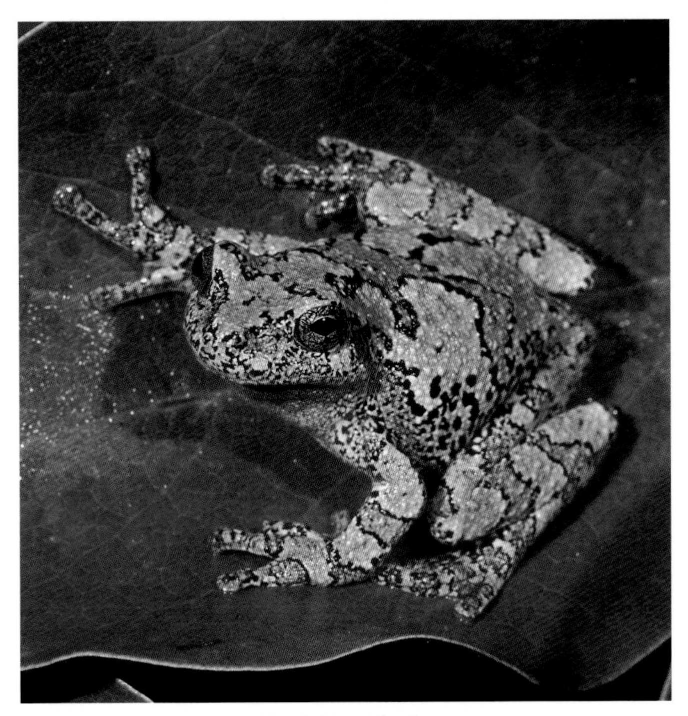

Cope's Gray Treefrog

The Making of the Book

The idea for this book took root in Lang Elliott's mind in the early 1990s, when he noticed that there weren't any books with audio compact discs that allowed people to enjoy beautiful photographs of frogs and toads, read about their natural histories, and hear their sounds. So Lang set about collecting recordings and photographs for what he presumed would be a long-term project. Several years later, Carl Gerhardt, an evolutionary biologist and frog-vocalization expert at the University of Missouri, joined in the effort, adding his own photographs and sound recordings to the pot.

In the late 1990s, Lang approached Houghton Mifflin Company with the idea of doing a book about the fifty or so species of frogs and toads found east of the Rocky Mountains. Houghton Mifflin liked this idea but suggested that the book be expanded to include all of North America. Because Lang and Carl had limited experience with western frogs and toads, Carlos Davidson was invited to join the team. Then a conservation ecologist at Sacramento State University, Carlos had already published two guides to the calls of western frogs and toads (see Sources and Further Reading, page 336).

With the three of us committed to the project, plans for the book finally began to take shape. But in spite of our combined resources, it became clear that we needed more materials. So we set about obtaining additional photographs and recordings, especially of western species, from a number of individuals (these contributors are acknowledged in the back of the book). For photographs, we are particularly indebted to Dick Bartlett of Gainesville, Florida, and Joe and Suzanne Collins of Lawrence, Kansas. Dick is the author of several books on amphibians, and Joe is the coauthor of the *Peterson Field Guide to Reptiles and Amphibians of Eastern and Central North America.* With their generous help, we finally had a remarkable variety of images upon which to draw.

The thrust to finish the book came in 2007. With deadlines approaching, we jumped into action, each of us playing a different role. Carl Gerhardt provided dozens of sound recordings and photographs and acted as the main writer for the book, producing most of the species accounts and family introductions, as well as the extended section on natural history. Carlos Davidson provided sound recordings for a number of western species, secured additional recordings from individuals and researchers, wrote the section on conservation, and reviewed and edited the western species accounts. Lang Elliott acted as team leader, overseeing the project from beginning to end. He edited the text, wrote a number of sections and species accounts, worked with reviewers, secured additional photographs and recordings, drew range maps, mixed and mastered the audio CD, developed and refined the design and layout of the book, and acted as the liaison with editors at Houghton Mifflin.

While the efforts of the authors and other contributors provided the essential ingredients, this book could not have become a reality if it were not for the enthusiastic crew at Houghton Mifflin Company. We applaud their steady commitment to publishing high-quality books about natural history. Our editor Lisa White was a joy to work with, and Anne Chalmers, designer extraordinaire, offered excellent and useful feedback throughout. We thank you, Houghton Mifflin, for making this project possible.

A "knot" of American Toads (five males trying to mount a lone female)

Acknowledgments

Many people have contributed to the creation of this book and we offer sincere thanks to all who have supported and encouraged us along the way, including but not limited to the following.

Special Thanks

We offer special thanks to the following individuals who have helped make this book beautiful and informative:

• Joe and Suzanne Collins of the Center for North American Herpetology (www.cnah.org), for supporting this project from the beginning, when it germinated nearly a decade ago, and for providing dozens of beautiful photographs and vital technical advice.

• Dick and Patti Bartlett, for providing continual support throughout the project, for donating dozens of beautiful photographs, for help in the field, and for contributing to the section on finding, observing, catching, and keeping frogs and toads.

• Michael Lannoo, Robert C. Stebbins, Joseph T. Collins, and Roger Conant, for their excellent field guides and reference works on North American frogs and toads (see Sources and Further Reading, page 336). We drew upon their works repeatedly in the creation of this book.

Our Academic Reviewers

The following herpetologists helped review and edit our species accounts:

Joseph T. Collins, Mark Hansen, Marc P. Hayes, Michael J. Lannoo, Emily Moriarty Lemmon, Paul Moler, and James C. Rorabaugh.

Lang Elliott's Acknowledgments

• Catherine Landis, my trusted companion who deeply shares my joy and appreciation of frogs and toads.

• The late Dean Metter, who not only taught me about the biology of frogs and toads, but who also led me into marshes and ponds to find them, when I was an undergraduate at the University of Missouri.

• All those who have accompanied me in the field or pointed me to good locations, including Paul Moler, John MacGregor, Walt Knapp, Keith Coleman, Wil Hershberger, Andy Price, Dick Bartlett, Mike Redmer, Dirk Stevenson, Julie Zickefoose, Jim White, Alvin Braswell, John Jensen, and many others.

• Beth Bannister for drawing range maps, sharing time in the field, and being her enthusiastic self.

• Lisa White and Anne Chalmers of Houghton Mifflin Company, for expert editing and helping me refine the design of the book.

Carl Gerhardt's Acknowledgments

• My wife, Dayna Glanz, who has often accompanied me in the field and has been supportive in so many ways.

• Richard DesRosiers, who introduced me to amphibians in general and frogs in particular during my formative years in Savannah, Georgia.

• Numerous students and colleagues at the University of Missouri, and especially Richard Daniel and the late Dean Metter, who have enjoyed observing frogs and toads in nature and not just as experimental animals.

• All those who have accompanied me in the field or pointed me to good locations, including Steve Bennett, Richard Daniel, Andy Price, Mike Sredl, and Brian Sullivan. Randy Babb was especially generous in sharing his deep knowledge of the amphibians and other wildlife in Arizona and in providing logistical help of all sorts.

Carlos Davidson's Acknowledgments

• My wife, Cynthia Kaufman, who has helped and accompanied me on many a frog trip.

• Pat Manley, who gave me the opportunity to make my first frog call CD, and to Brad Shaffer for giving me the opportunity to make studying frogs a career.

• The many people who helped me find recording sites, including Joe DiDonata and Roger Epperson of the East Bay Regional Parks District, Brian Sullivan, Mark Jennings, Marc Hayes, M. J. Foquette, Phil Fernandez, Jim Collins, Terry Myers, Brad Shaffer, John Gerhart, Gary M. Fellers, Jim Rorabaugh, and Michelle Barlow.

Thank You, Frogs and Toads

Last but not least, we applaud the frogs and toads themselves, who have given us the opportunity to celebrate their lives. We trust that humankind will keep their habitats healthy and diverse, so that they forever continue to delight us with their leaping, splashing, and calling.

Pine Woods Treefrogs

The Photographs

A total of 388 extraordinary photographs by 27 photographers grace the pages of this book. We are delighted with the quality throughout, and we believe that our book will be valued in large measure by the beauty of its images. Below is a complete list of photographers, arranged according to the number of images that each provided. Photographs are identified by page number. Small inset photos are designated by an "i" after the page number. On pages containing more than one photo, the photos are listed from top to bottom, left to right, or clockwise from the upper left using the letters A through D.

Lang Elliott—front cover, back cover, 1, 2, 3, 5, 6, 9, 10, 11, 14A, 14B, 14C, 14D, 15A, 15E, 19, 20, 29, 36, 42, 44, 45, 47, 48A, 48B, 49, 51, 52, 53, 55, 57, 58, 59A, 59B, 59D, 60A, 60B, 61, 61i, 62, 64, 65A, 65C, 65D, 67, 68B, 70B, 76, 77, 78, 79, 81, 83, 84B, 93, 96A, 96B, 101, 102, 103, 103i, 104, 105, 106, 107, 114A, 114B, 116B, 116D, 118, 119, 120, 121, 124A, 127, 128, 129, 130, 131, 132, 133, 135, 138, 139, 140, 141, 143, 144A, 144B, 145, 146, 147, 148, 149, 154, 161, 185, 186, 187, 188, 189, 190, 191, 192, 193, 194A, 194B, 195, 198, 199, 200, 201, 203, 205, 206, 207, 214, 215, 217, 218, 219A, 219B, 220, 221, 223, 225, 226, 227, 254, 256, 257, 257i, 260, 261, 263, 264A, 267, 276, 277, 278, 279, 281, 287, 290, 291, 297, 311, 315, 317, 319, 320, 323, 326, 331, 334, 341, 343, 344

Suzanne L. Collins, CNAH—15D, 33A, 37, 75A, 82B, 86, 87A, 87B, 90B, 91, 92, 95, 97, 98, 110B, 113A, 116A, 125, 134, 136, 137, 142A, 150, 151, 152, 157, 159, 165A, 165B, 167, 169, 174, 197, 204B, 211, 213, 237, 241, 247, 249, 264B, 268, 271, 274, 283, 288B, 289, 294, 296A

Dick Bartlett—15C, 21A, 21B, 31, 39, 50, 56, 59C, 73i, 80, 82A, 85, 99, 117, 122, 124B, 142B, 158A, 171, 172, 175A, 175B, 184, 196, 202, 204A, 209, 210, 212, 219C, 219D, 238B, 258, 259, 264D, 269, 275, 285, 286A, 286B, 288A, 296B, 298, 299, 300A, 302, 304B, 309

Carl Gerhardt—24, 27, 30A, 30B, 46, 54, 63, 65B, 69, 70A, 71, 72, 73, 74, 75B, 84A, 100, 113C, 123, 123i, 168A, 222, 224, 230, 231, 232A, 232B, 233, 234, 239, 266B, 301, 304A

Robert Wayne Van Devender—33B, 110D, 112, 156, 158B, 236, 242B, 273, 284, 292, 293, 295, 300B, 308

Erik Enderson—16, 160A, 160B, 163, 164, 166, 170A, 170B, 216, 266A, 270, 280

Ornate Chorus Frog

William Leonard—15B, 111, 162, 173, 177, 183i, 236B, 240, 244, 246, 303, 312

Gary Nafis—34, 125i, 229, 238A, 242A, 248A, 250, 262

Robert English, LEAPS—23, 43, 66, 68A, 115, 116C, 321, 335

Chris W. Brown—173i, 181, 252, 253, 272, 305

Robert H. Hansen—176, 179, 180, 183, 243, 251

John Cancalosi—12, 155, 255, 264C, 265

Toby J. Hibbitts—282A, 282B, 306A, 306B, 307

Joshua L. Puhn—108, 110A, 110C, 182, 248B

William C. Flaxington—113B, 178

Wil Hershberger—89, 337

Tom R. Johnson—153, 208

Alan St. John—109, 245

John White—88, 89i

Roger J. Allis—199i

Randall D. Babb—235

Tony Gamble—94

Joyce Gross—25

Randy D. Jennings—228

James C. Rorabaugh—168B

Scott A. Smith—91i

Jim White—90A

Barking Treefrog

The Sound Recordings

The audio compact disc that accompanies this book features the recordings of nearly every species of frog and toad found in North America, north of Mexico. The only exceptions are the two tailed frogs, *Ascaphus truei* and *Ascaphus montanus,* neither of which are known to make vocal sounds. Twenty-seven individuals and four institutions contributed a total of 196 recordings to the project. A complete documentation of the recordings used to construct each track can be found on the following pages. Below is a summary list of contributors—the number of recordings from each source is indicated after each name:

Individuals:

Lang Elliott (119)
Carl Gerhardt (18)
Carlos Davidson (13)
Gary Nafis (7)
Cynthia K. Sherman (3)
Timothy C. Ziesmer (3)
Barb Beck (2)
Emily Moriarty Lemmon (2)
Ted Mack (2)
John Neville (2)
Philip T. Northern (2)
Brian K. Sullivan (2)
Frank T. Awbrey (1)
Jeffrey L. Briggs (1)
Wyoming Fish and Game (1)
John Hartog (1)
Randy D. Jennings (1)
John Jensen (1)
Walt Knapp (1)
Alejandro Purgue (1)
Richard Siegel and Steven Richter (1)
Martyn Stewart (1)
Stan Tekiela (1)
Michael L. Treglia (1)
Eric Wallace (1)
Jim White (1)

Institutions:

Macaulay Library, Cornell University Laboratory of Ornithology:
- Jonathon Storm (1)
- Greg F. Budney and Karin Hoff (1)
- Theodore A. Parker III (1)

Borror Laboratory of Bioacoustics:
- Erik D. Lindquist (1)

Texas Natural Science Center:
- W. Frank Blair (1)

Florida Museum of Natural History:
- David Lee (1)

Compact Disc Track Descriptions

Treefrogs and Allies—Hylidae

Hyla—Treefrogs

1. Green Treefrog—*Hyla cinerea*
- Small chorus. May 8, 1988, 10:30 pm, St. Marks National Wildlife Refuge, Florida. Lang Elliott.
- Large chorus. May 2, 1994, before midnight, Wacissa River near Newport, Florida. Lang Elliott.
- Interaction with aggressive calls. May 19, 1994, 11:30 pm, Wacissa River, near Newport, Florida. Ted Mack.

2. Barking Treefrog—*Hyla gratiosa*
- Small chorus. March 23, 1989, 9:10 pm, 62°F, Ocala National Forest, Florida. Lang Elliott.
- Large chorus. June 10, 1997, 10 pm, Ocala National Forest, Florida. Lang Elliott.

3. Pine Barrens Treefrog—*Hyla andersonii*
- Close-up of male. May 10, 1992, 10:30 pm, Carolina Sandhills National Wildlife Refuge, South Carolina. Lang Elliott.
- Two males calling. June 10, 1994, 11:55 pm, 70°F, Carolina Sandhills National Wildlife Refuge, South Carolina. Lang Elliott.

4. Squirrel Treefrog—*Hyla squirella*
- Small chorus; one male close. May 30, 1995, 11:45 pm, 72°F, Appalachicola National Forest, Florida. Lang Elliott.
- Single male calling. May 1, 1994, 8:15 pm, 76°F, Big Cypress National Preserve, Florida. Lang Elliott.
- Rain call given from shrub. April 30, 1989, 5 am, Tall Timbers Research Station, near Tallahassee, Florida. Lang Elliott.

5. Pine Woods Treefrog—*Hyla femoralis*
- Two males calling. April 28, 1994, 11 pm, 70°F, Appalachicola National Forest, Florida. Lang Elliott.
- Several males calling. June 5, 1994, 12:30 am, 76°F, Kissimmee Prairie Audubon Refuge, near Basinger, Florida. Lang Elliott.

6. Gray Treefrog—*Hyla versicolor*
- Small chorus. May 22, 1992, 1 am, Connecticut Hill Wildlife Management Area near Ithaca, New York. Lang Elliott.
- Small chorus. May 18, 1994, 9:30 pm, 78°F, Connecticut Hill Wildlife Management Area, near Ithaca, New York. Lang Elliott.
- Two males alternating calls. May 7, 2007, 9:54 pm, 66°F, near Columbia, Missouri. Lang Elliott.
- Interaction with aggressive calls. May 20, 1986, 10:30 pm, Ithaca, New York. Lang Elliott.

7. Cope's Gray Treefrog—*Hyla chrysoscelis*
- Close-up of male. May 14, 2001, 10 pm, 68°F, Land Between the Lakes, Kentucky. Lang Elliott.
- Close-up of male. May 16, 1988, 8:30 pm, 70°F, Land Between the Lakes, Kentucky. Lang Elliott.
- Close-up of male. April 26, 1991, 72°F, Jasper County, South Carolina. Carl Gerhardt.

8. Bird-voiced Treefrog—*Hyla avivoca*
- Close-up of male then aggressive calls and call of second male. June 10, 2001, 10:30 pm, 75°F, Appalachicola National Forest, Florida. Lang Elliott.

9. Arizona Treefrog—*Hyla wrightorum*
- Two males alternating calls; one close. August 8, 2003, 62°F, near Payson, Arizona. Carl Gerhardt.
- Two males alternating calls. July 22, 2007, 12:35 am, 62°F, near Forest Lakes, Arizona. Carl Gerhardt.

10. Canyon Treefrog—*Hyla arenicolor*
- Calls of warm males. May 6, 2004, 72°F, Santa Rita Mountains, near Greaterville, Arizona. Carl Gerhardt.
- Calls of cool males. 2004, 56°F, Sierra Anchas Mountains, near Fulton, Arizona. Carl Gerhardt.

- Aggressive call. May 16, 2005, 68°F, Diamond Creek, Grand Canyon, Arizona. Carl Gerhardt.

Pseudacris—Chorus Frogs

11. Spring Peeper—*Pseudacris crucifer*
- Two males alternating calls. April 13, 1992, 9:30 pm, Connecticut Hill Wildlife Management Area, near Ithaca, New York. Lang Elliott.
- Two males calling. April 26, 1994, 11:45 pm, Land Between the Lakes, Kentucky. Lang Elliott.
- Small Chorus. May 14, 1990, 9:30 pm, Shindagin Hollow, near Ithaca, New York. Lang Elliott.
- Aggressive calls. April 24, 1990, 9 pm, Connecticut Hill Wildlife Management Area near Ithaca, New York. Lang Elliott.
- Rain calls given from bush. September 15, 1991, 6 pm, Connecticut Hill Wildlife Management Area near Ithaca, New York. Lang Elliott.

12. Little Grass Frog—*Pseudacris ocularis*
- Close-up of calling male. June 1, 1994, 7:30 pm, 78°F, Big Cypress National Preserve, Florida. Lang Elliott.
- Several males calling. May 5, 1994, 12:05 am, 78°F, Kissimmee Prairie Audubon Refuge, near Basinger, Florida. Lang Elliott.

13. Ornate Chorus Frog—*Pseudacris ornata*
- Single male calling. March 17, 2001, 10 pm, 53°F, near Butler, Georgia. Lang Elliott.
- Two males alternating calls. March 17, 2001, 11:45 pm, 51°F, near Butler, Georgia. Lang Elliott.
- Small chorus. March 21, 2001, 9 pm, 48°F, Blackwater State Park, near Munson, Florida. Lang Elliott.

14. Strecker's Chorus Frog—*Pseudacris streckeri*
- Small chorus. March 30, 2001, 11:30 pm, 51°F, near Willow City, Texas. Lang Elliott.

15. Midland Chorus Frog—*Pseudacris triseriata*
- Two males alternating calls. April 26, 2008, 52°F, Pigeon River Fish and Wildlife Area, near Mongo, Indiana. Emily Moriarty Lemmon.

16. Upland Chorus Frog—*Pseudacris feriarum*
- Small chorus; one male close. March 14, 2002, 55°F, Obion County, Tennessee. Emily Moriarty Lemmon.

17. New Jersey Chorus Frog—*Pseudacris kalmi*
- Two males alternating calls. April 13, 2002, 65°F, Kent County, Maryland. Emily Moriarty Lemmon.
- Small chorus. February 2000. Bombay Hook National Wildlife Area, near Smyrna, Delaware. Jim White.

18. Cajun Chorus Frog—*Pseudacris fouquettei*
- Two males alternating calls. March 22, 2001, 9:15 pm, 59°F, Felsenthal National Wildlife Refuge, Arkansas. Lang Elliott.
- Several males. March 22, 2001, 10 pm, 59°F, Felsenthal National Wildlife Refuge, Arkansas. Lang Elliott.

19. Boreal Chorus Frog—*Pseudacris maculata*
- Two males alternating calls. May 4, 1993, 11:57 pm, 45°F, Riding Mountain National Park, Manitoba. Lang Elliott.
- Chorus. May 30, 1993, Riding Mountain National Park, Manitoba. Lang Elliott.

20. Southern Chorus Frog—*Pseudacris nigrita*
- Small chorus. March 14, 2001, 62°F, Fort Stewart, Georgia. Lang Elliott.
- Close-up of warm male. May 8, 1997, 9:30 pm, about 78°F, Everglades National Park, Florida. Lang Elliott.

21. Brimley's Chorus Frog—*Pseudacris brimleyi*
- Two males alternating calls. March 11, 2001, 3 am, 43°F, near Maysville, North Carolina. Lang Elliott.
- Two males alternating calls. March 12, 2001, 1 am, 61°F, near Maysville, North Carolina. Lang Elliott.

22. Mountain Chorus Frog—*Pseudacris brachyphona*
- Close-up of male. April 18, 2003, 8 pm, 61°F, Big South Fork National Recreation Area, Tennessee. Lang Elliott.
- Small chorus. May 1, 1996, 9:30 pm, near Marietta, Ohio. Lang Elliott.
- Two males alternating calls. April 17, 2003, 9 pm, 60°F, Mammoth Cave National Park, Kentucky. Lang Elliott.

23. Spotted Chorus Frog—*Pseudacris clarkii*

- Close-up of male. April 15, 2001, 1 am, 66°F, Waurika Wildlife Management Area, east of Walters, Oklahoma. Lang Elliott.
- Two males alternating calls. April 15, 2001, 1:30 am, 66°F, Waurika Wildlife Management Area, east of Walters, Oklahoma. Lang Elliott.
- Small chorus. April 7, 2001, 11 pm, 52°F, east of Silverton, Texas. Lang Elliott.

24. Pacific Chorus Frog—*Pseudacris regilla*

- Typical two-parted call of male. May 22, 1994, 47°F, San Bernardino National Forest, Riverside County, California. Carlos Davidson.
- One-syllable call variant. April 12, 1991, 66°F, Cotati, Sonoma County, California. Philip. T. Northern.
- Calls of cold males. March 7, 2002, 45°F, near Corvallis, Oregon. Carl Gerhardt.
- Dry land call. January 23, 1994, 60°F, Santa Cruz Island, Santa Barbara County, California. Carlos Davidson.
- Dry land call. November 5, 2006. Salt Spring Island, British Columbia. John Neville.

25. California Chorus Frog—*Pseudacris cadaverina*

- Two males alternating calls, with aggressive call. April 11, 2007, 61°F, Santa Margarita Ecological Preserve, Deluz, California. Carl Gerhardt.
- Aggressive calls. April 11, 2007, 61°F, Santa Margarita Ecological Preserve, Deluz, California. Carl Gerhardt.

Acris—Cricket Frogs

26. Northern Cricket Frog—*Acris crepitans*

- Call sequence of close male. May 12, 2001, 11:30 pm, 65°F, Lost Maples State Natural Area, near Vanderpool, Texas. Lang Elliott.
- Small group. May 28, 1988, 10:30 pm, 65°F, Mingo National Wildlife Refuge, near Puxico, Missouri. Lang Elliott.
- Large chorus. May 1, 1991, 9:30 pm, Land Between the Lakes, Kentucky. Lang Elliott.

27. Southern Cricket Frog—*Acris gryllus*

- Call sequence of close male. April 27, 2005, 9:15 pm, 69°F, St. Marks National Wildlife Refuge, near Newport, Florida. Lang Elliott.
- Small group; one close male. July 11, 1989, 12:25 am, Carolina Sandhills National Wildlife Refuge, South Carolina. Lang Elliott.
- Large chorus. April 28, 1994, 11:50 pm, 70°F, Appalachicola National Forest, Florida. Lang Elliott.

Osteopilus—West Indian Treefrogs

28. Cuban Treefrog—*Osteopilus septentrionalis*

- Several males; one close. May 2, 1994, 2:30 am, 79°F, Everglades National Park, Florida. Lang Elliott.
- Large chorus. May 2, 1994, 2:15 am, 79°F, Everglades National Park, Florida. Lang Elliott.

Smilisca—Mexican Treefrogs

29. Lowland Burrowing Treefrog—*Smilisca fodiens*

- Two males alternating calls. August 5, 2003, Vekol Valley, Arizona. Carl Gerhardt.
- Small chorus with alternating calls. August 5, 2003, Vekol Valley, Arizona. Carl Gerhardt.

30. Mexican Treefrog—*Smilisca baudinii*

- Close-up of male, with chuckling notes. July 27, 2008, 1 am, 80°F, Brownsville, Texas. Lang Elliott.
- Small chorus; one male close. July 25, 2008, 10:30 pm, 82°F, near San Benito, Texas. Lang Elliott.

True Toads—Bufonidae

Bufo—True Toads

31. American Toad—*Bufo americanus*

- Trills of two close males. May 18, 1989, 11 pm, 64°F, near Ithaca, New York. Lang Elliott.
- Release calls of handheld male. May 18, 1989, 10 pm, 64°F, near Ithaca, New York. Lang Elliott.

32. Fowler's Toad—*Bufo fowleri*
- Calls of lone male. May 14, 2001, 10 pm, 68°F, Land Between the Lakes, Kentucky. Lang Elliott.
- Calls of male. May 14, 2001, 10:15 pm, 68°F, Land Between the Lakes, Kentucky. Lang Elliott.
- Chorus. May 30, 1991, 10 pm, Land Between the Lakes, Kentucky. Lang Elliott.

33. Woodhouse's Toad—*Bufo woodhousii*
- Calls of several males; one close. May 11, 1993, 11 pm, Theodore Roosevelt National Park, North Unit, North Dakota. Lang Elliott.

34. Southern Toad—*Bufo terrestris*
- Small group. May 3, 1992, 11:40 pm, Osceola National Forest, Florida. Lang Elliott.
- Large chorus. May 31, 1994, 1:09 am, 72°F, Appalachicola National Forest, Florida. Lang Elliott.

35. Oak Toad—*Bufo quercicus*
- Calls of close male. June 1, 1994, 7:45 pm, 77°F, Big Cypress National Preserve, Florida. Lang Elliott.
- Small group. May 1, 1994, 7:50 pm, 77°F, Big Cypress National Preserve, Florida. Lang Elliott.

36. Coastal Plain Toad—*Bufo nebulifer*
- Calls of close male. April 1, 2001, 2 am, 66°F, Beuscher State Park, near Bastrop, Texas. Lang Elliott.
- Small group. April 2, 2001, 67°F, Beuscher State Park near Bastrop, Texas. Lang Elliott.

37. Texas Toad—*Bufo speciosus*
- Calls of close male. April 15, 2001, 4:30 am, 63°F, Hackberry Flat Wildlife Management area, near Frederick, Oklahoma. Lang Elliott.
- Release calls of captured male. April 17, 2001, Hackberry Flat Wildlife Management Area, near Frederick, Oklahoma. Lang Elliott.

38. Houston Toad—*Bufo houstonensis*
- Calls of two males. April 23, 2001, 2:30 am, 60°F, Bastrop State Park, near Bastrop, Texas. Lang Elliott.

39. Cane Toad—*Bufo marinus*
- Close-up of male. July 26, 2008, 1:30 am, 80°F, south of Weslaco, Texas. Lang Elliott.
- Two males singing. July 26, 2008, 1 am, 80°F, south of Weslaco, Texas. Lang Elliott.

40. Great Plains Toad—*Bufo cognatus*
- Call of one male. May 5, 2001, 2 am, 62°F, Hackberry Flat Wildlife Management Area, near Frederick, Oklahoma. Lang Elliott.
- Call of one male. April 21, 2001, 11 pm, 60°F, Franklin Bottoms, near Booneville, Missouri. Lang Elliott.

41. Canadian Toad—*Bufo hemiophrys*
- Call of single male. Early June, 10 pm, 45°F, Wood Buffalo National Park, Alberta, Canada. John Neville.
- Several males calling. Recorded near Edmonton, Alberta. Barb Beck.

42. Wyoming Toad—*Bufo baxteri*
- Calls of one male. May 15, 1989, 11:30 pm, 66°F, Mortenson Lake, Adams County, Wyoming. Greg Hallen and Don R. Miller (Wyoming Fish and Game).

43. Red-spotted Toad—*Bufo punctatus*
- Calls of one male. May 9, 2001, 11 pm, 63°F, Lost Maples State Natural Area, near Vanderpool, Texas. Lang Elliott.

44. Green Toad—*Bufo debilis*
- Small chorus; one male close. August 8, 2003, 74°F, near McNeal, Arizona. Carl Gerhardt.

45. Sonoran Green Toad—*Bufo retiformis*
- Calls of one male. August 5, 2003, Vekol Valley, Arizona. Carl Gerhardt.

46. Sonoran Desert Toad—*Bufo alvarius*
- Small chorus; one male close. August 24, 1992, 79°F, Sullivan Skunk Creek, Maricopa County, Arizona. Brian K. Sullivan.

47. Arizona Toad — *Bufo microscaphus*

- Small chorus; one male close. March 4, 1994, 60°F, Hassayampa River, near Wickenberg, Arizona. Brian K. Sullivan.

48. Western Toad — *Bufo boreas*

- Two males calling *(B. b. boreas)*. Recorded near Edmonton, Alberta. Barb Beck.
- Call series of one male *(B. b. boreas)*. Recorded near Edmonton, Alberta. Barb Beck.
- Calls of one male *(B. b. halophilus)*. May 24, 1973, 72°F, Cleveland National Forest, near Descanso, California. Frank T. Awbrey.

49. Amargosa Toad — *Bufo nelsoni*

- Release calls of captive male. February 28, 1995. Recorded in laboratory at the University of Nevada. Gregory F. Budney and Karin Hoff (Macaulay Library).

50. Black Toad — *Bufo exsul*

- Small group giving release calls. March 28, 1978, Corral Springs, Deep Springs Valley, Inyo, California. Cynthia K. Sherman.

51. Arroyo Toad — *Bufo californicus*

- Small chorus; two males close. May 20, 1994, 66°F, Los Padres National Forest, Santa Barbara County, California. Carlos Davidson.
- Release calls of handheld individual. May 20, 1994, 66°F, Los Padres National Forest, Santa Barbara County, California. Carlos Davidson.

52. Yosemite Toad — *Bufo canorus*

- Small chorus; one male close. May 3, 1977, 70°F, Inyo National Forest, Mono County, California. Cynthia K. Sherman.
- Small chorus. May 3, 1977, 70°F, Inyo National Forest, Mono County, California. Cynthia K. Sherman.

True Frogs — Ranidae

Rana — True Frogs

53. Bullfrog — *Rana catesbeiana*

- Small chorus. May 7, 1989, 3 am, Michigan Hollow, near Ithaca, New York. Lang Elliott.
- Aggressive spit calls. April 20, 1989, 9 pm, Felsenthal National Wildlife Refuge, Arkansas. Lang Elliott.
- Alarm squeaks and splashes. September 2, 1994, 75°F, Point Reyes National Seashore, Marin County, California. Carlos Davidson and Cynthia Kaufman.

54. Green Frog — *Rana clamitans*

- Small chorus. July 28, 1988, 3 am, 70°F, Connecticut Hill Wildlife Management Area, near Ithaca, New York. Lang Elliott.
- Stuttered calls of male. May 27, 1987, Connecticut Hill Wildlife Management Area, near Ithaca, New York. Lang Elliott.
- Typical calls of Bronze Frog subspecies. May 7, 2001, 11:30 pm, 80°F, Eglin Air Force Base, near Crestview, Florida. Lang Elliott.
- Alarm squeaks and splashes. September 15, 1991, 6:30 pm, Connecticut Hill Wildlife Management Area, near Ithaca, New York. Lang Elliott.

55. Pig Frog — *Rana grylio*

- Small chorus; one male close. April 22, 1989, 1 am, Sabine National Wildlife Refuge, Louisiana. Lang Elliott.
- Chorus. May 11, 1994, 11 pm, Okeefenokee National Wildlife Refuge, Georgia. Lang Elliott.
- Aggressive spit calls. April 26, 1989, 8:30 pm, 75°F, St. Marks National Wildlife Refuge, Florida. Lang Elliott.
- Distress calls of handheld individual. May 12, 1994, 5:30 am, Okeefenokee National Wildlife Refuge, Georgia. Lang Elliott.

56. Mink Frog — *Rana septentrionalis*

- Small chorus; one male close. May 24, 1989, 3 am, Adirondack Mountains, near Loon Lake, New York. Lang Elliott.

- Large chorus. July 4, 1991, 3 am, Adirondack Mountains, near Paul Smiths, New York. Lang Elliott.

57. Wood Frog—*Rana sylvatica*
- Small chorus. March 30, 1988, 11:30 pm, 35°F, Connecticut Hill Wildlife Management Area, near Ithaca, New York. Lang Elliott.
- Large chorus. April 1, 2004, 12:30 am, 42°F, Connecticut Hill Wildlife Management Area, near Ithaca, New York. Lang Elliott.

58. Carpenter Frog—*Rana virgatipes*
- Small chorus; one male close. March 30, 1989, 9:45 pm, 67°F, Cheraw State Park, Cheraw, South Carolina. Lang Elliott.
- Chorus. May 3, 1996, Wharton State Forest, New Jersey Pine Barrens, New Jersey. Ted Mack.

59. Florida Bog Frog—*Rana okaloosae*
- Two males interacting. June 7, 2001, 12:30 am, 75°F, Eglin Air Force Base, near Crestview, Florida. Lang Elliott.
- Calls of one male. May 1, 1992, 11:30 pm, 58°F, Eglin Air Force Base, near Crestview, Florida. Lang Elliott.

60. River Frog—*Rana heckscheri*
- Small chorus ; one male close. May 7, 1994, 1:30 am, 76°F, Osceola National Forest, Florida. Lang Elliott.

61. Crawfish Frog—*Rana areolata*
- Small chorus. March 23, 2007, 8:20 pm, 60°F, Caldwell County, Kentucky. Lang Elliott.
- Large chorus. March 21, 2007, 10 pm, 62°F, Caldwell County, Kentucky. Lang Elliott.

62. Gopher Frog—*Rana capito*
- Two males calling. February 20, 1998, Okaloosa County, Florida. John Jensen and Lee Andrews.
- Small chorus. February 21, 1975, Putnam County, Florida. David Lee (Florida Museum of Natural History).

63. Dusky Gopher Frog—*Rana sevosa*
- Small chorus. March 8, 1988, Desoto National Forest, Harrison County, Mississippi. Rich Siegel and Steve Richter.

64. Pickerel Frog—*Rana palustris*
- Small chorus. April 22, 2006, 12:30 am, 58°F, near Staunton, Virginia. Lang Elliott.
- Staccato note, snores, and garbled call of male. April 27, 1990, 11 pm, Connecticut Hill Wildlife Management Area, near Ithaca, New York. Lang Elliott.

65. Northern Leopard Frog—*Rana pipiens*
- Snores and grunts of close male. April 29, 2001, 11:30 pm, 69°F, Mile Lac State Wildlife Management Area, near Milaca, Minnesota. Lang Elliott.

66. Southern Leopard Frog—*Rana sphenocephala*
- Two males alternating calls. March 4, 1989, 4:30 am, DeSoto National Forest, Mississippi. Lang Elliott.
- Small chorus; one male close. May 5, 1994, 1:30 am, 76°F, Kissimmee Prairie Audubon Reserve, near Basinger, Florida. Lang Elliott.
- Excited chorus. May 3, 2001, 2:45 am, 75°F, Appalachicola National Forest, near Crawfordsville, Florida. Lang Elliott.

67. Plains Leopard Frog—*Rana blairi*
- Small chorus; one male close. May 24, 1994, 10 pm, 72°F, Quivira National Wildlife Refuge, Kansas. Lang Elliott.
- Small chorus. April 26, 2001, 11 pm, 65°F, near Overbrook, Kansas. Lang Elliott with Keith Coleman.

68. Rio Grande Leopard Frog—*Rana berlandieri*
- Intermittent calls of male. May 12, 2001, 11:30 pm, 65°F, Lost Maples State Natural Area, near Vanderpool, Texas. Lang Elliott.
- Small chorus. May 11, 2001, 1:17 am, 71°F, Chaparral Wildlife Management Area, near Artesia Wells, Texas. Lang Elliott.

69. Relict Leopard Frog—*Rana onca*

- Chuckles and scrapes of male. February 28, 1992, 66°F, Lake Mead National Recreation Area, Clark County, Nevada. Randy D. Jennings.

70. Chiricahua Leopard Frog—*Rana chiricahuensis*

- Scrapes and chucks of single male. July 13, 2003, 61°F, Three Forks, Navajo County, Arizona. Carl Gerhardt.
- Small chorus. August 19, 1994, 8:20 pm, 56°F, White Mountains, Apache-Sitgreaves National Forest, Apache County, Arizona. Carlos Davidson.

71. Lowland Leopard Frog—*Rana yavapaiensis*

- Chorus. May 20, 2002, 70°F (water), Muleshoe Ranch in the Galiuro Mountains, Cochise County, Arizona. Carl Gerhardt.

72. Tarahumara Frog—*Rana tarahumarae*

- Several individuals at an artificial breeding tank. July 24, 2007, Castle Dome Mountains, Yuma County, Arizona. Carl Gerhardt.

73. Cascades Frog—*Rana cascadae*

- Chuckles of single male. July 7, 2006, 2 pm, 70°F, Mt. Ranier National Park, Washington. Gary Nafis.
- Chatter and other calls given during encounter between two males. July 7, 2006, 2 pm, 70°F, Mt. Ranier National Park, Washington. Gary Nafis.
- Excited calls of two males. March 9, 1968, 54°F, Bear Creek, Jefferson County, Oregon. Jeffrey L. Briggs.

74. Northern Red-legged Frog—*Rana aurora*

- Males calling at surface of water. February 29, 1992, 68°F, Fresh Water Lagoon, Del Norte County, California. Alejandro Purgue.
- Underwater calls. February 7, 2006, 3:30 pm, 50°F, Lewis County, Washington. Gary Nafis.

75. California Red-legged Frog—*Rana draytonii*

- Small chorus. March 2, 2007. Morgan Territory Regional Preserve, Contra Costa County, California. Carlos Davidson.
- Chuckles and grunts of male. January 31, 2007. Calavera Creek, near Pacifica, California. Carlos Davidson.

76. Oregon Spotted Frog—*Rana pretiosa*

- Call series of male. Feruary 24, 2006, 2:30 pm, 45°F, Thurston County, Washington. Gary Nafis.

77. Columbia Spotted Frog—*Rana luteiventris*

- Call series of male. March 17, 2002, 9:50 pm, 42°F, Mudflat Road, eastern Washington. Martyn Stewart.
- Call series of male. April 25, 1994, 54°F, Okanogan National Forest, Okanogan County, near Winthrop, Washington. Jonathon Storm.

78. Foothill Yellow-legged Frog—*Rana boylii*

- Calls of one male and soft squeaks of another. May 7, 1993, 59°F, Pepperwood Ranch, Sonoma County, California. Tim C. Ziesmer.
- Underwater calls. June 26, 11 pm, 78°F, Linn County, Oregon. Gary Nafis.
- Underwater calls. June 27, 78°F, Linn County, Oregon. Gary Nafis.

79. Sierra Madre Yellow-legged Frog—*Rana muscosa*

- Calls of male. May 24, 1994, 65°F, Hall Research Natural Area, San Bernardino National Forest, Riverside County, California. Tim C. Ziesmer.

80. Sierra Nevada Yellow-legged Frog—*Rana sierrae*

- Calls of one male. July 1, 2006, 12:30 pm, near Ebbet's pass, Alpine County, California. Gary Nafis.
- Underwater calls. May 14, 1993, 65°F, Summit Meadow, Yosemite National Park, California. Tim C. Ziesmer.

North American Spadefoots—Scaphiopodidae

Scaphiopus—Southern Spadefoots

81. Eastern Spadefoot—*Scaphiopus holbrookii*

- Small chorus; one male close. March 12, 1973, Scott County, Missouri. Carl Gerhardt.
- Large chorus. March 5, 2003, 9 pm, near Herndon, Jenkins County, Georgia. Walt Knapp.

82. Hurter's Spadefoot—*Scaphiopus hurterii*
- Two males calling; one very close. April 1, 2001, 3 am, 66°F, Beuscher State Park, near Bastrop, Texas. Lang Elliott.
- Chorus. April 1, 2001, 2:30 am, 66°F, Beuscher State Park, near Bastrop, Texas. Lang Elliott.

83. Couch's Spadefoot—*Scaphiopus couchii*
- Several males calling. May 5, 2001, 12:30 am, 64°F, near Grandfield, Oklahoma. Lang Elliott.

Spea—Western Spadefoots

84. Plains Spadefoot—*Spea bombifrons*
- Calls of one male. May 5, 2001, 2:30 am, 62°F, Hackberry Flat Wildlife Management Area, near Frederick, Oklahoma. Lang Elliott.
- Two males alternating calls. May 5, 2001, 3 am, 62°F, Hackberry Flat Wildlife Management Area, near Frederick, Oklahoma. Lang Elliott.
- Release calls of handheld male. April 21, 2001, 70°F, Franklin Bottoms, near Booneville, Missouri. Lang Elliott.

85. Great Basin Spadefoot—*Spea intermontana*
- Chorus. April 29, 1968, 11:05 pm, 49°F, Veyo, Washington County, Utah. Philip T. Northern.

86. New Mexico Spadefoot—*Spea multiplicata*
- Two males alternating calls. 2003, near Heber, Arizona. Carl Gerhardt.
- Three males calling. August 23, 1994, 66°F, near Portal, Cochise County, Arizona. Carlos Davidson.

87. Western Spadefoot—*Spea hammondii*
- Two males calling; one very close. March 12, 1995, 9:15 pm, Coral Hollow, Alameda County, California. Carlos Davidson.
- Small chorus. March 12, 1995, 9:45 pm, Coral Hollow, Alameda County, California. Carlos Davidson.

Microhylid Frogs and Toads—Microhylidae

Gastrophryne—North American Narrowmouth Toads

88. Eastern Narrowmouth Toad—*Gastrophryne carolinensis*
- Several males calling. July 23, 1973, Chatham County, Georgia. Carl Gerhardt.
- Small chorus, with aggressive calls at end. May 30, 1994, 10 pm, 72°F, Appalachicola National Forest, Florida. Lang Elliott.

89. Great Plains Narrowmouth Toad—*Gastrophryne olivacea*
- Small chorus. May 6, 2001, 11 pm, 64°F, next to Wichita Mountains National Wildlife Refuge, near Lawton, Oklahoma. Lang Elliott.

Hypopachus—Sheep Frogs

90. Sheep Frog—*Hypopachus variolosus*
- Close-up of male. July 29, 2008, 12:30 am, 80°F, near Raymondville, Texas. Lang Elliott.

Neotropical Frogs and Toads—Leptodactylidae

Eleutherodactylus—Rain Frogs

91. Greenhouse Frog—*Eleutherodactylus planirostris*
- Calls of male. May 28, 2008, 5:25 am, Long Pine Key campground, Everglades National Park, Florida. Lang Elliott.
- Calls of male. May 30, 2008, 5:50 am, Big Cypress National Preserve, Florida. Lang Elliott.

92. Puerto Rican Coqui—*Eleutherodactylus coqui*
- Chorus; one male very close. April 28, 1988, Puerto Rico. Theodore A. Parker III (Macaulay Library).
- Chorus; one male close. February 11, 2004, 11:30 pm, Kalopa Forest Reserve, Big Island, Hawaii. John Hartog.

93. Cliff Chirping Frog—*Eleutherodactylus marnockii*
- Chirps and trills of male. April 5, 2001, 11 pm, 65°F, Lost Maples State Natural Area, near Vanderpool, Texas. Lang Elliott.

94. Spotted Chirping Frog—*Eleutherodactylus guttilatus*
- Chirps and trills of one male. July 29, 2001, near Alpine, Texas. Eric Wallace.

95. Rio Grande Chirping Frog—*Eleutherodactylus cystignathoides*
- Chirps and trills of one male. May 4, 2008, 10:30 pm, 75°F, San Benito, Texas. Lang Elliott.

Craugastor—Robber Frogs

96. Barking Frog—*Craugastor augusti*
- Calls of one male. April 4, 2001, 9 pm, 72°F, Lost Maples State Natural Area, near Vanderpool, Texas. Lang Elliott.
- Two males interacting. April 4, 2001, 9:45 pm, 70°F, Lost Maples State Natural Area, near Vanderpool, Texas. Lang Elliott.

Leptodactylus—Neotropical Grass Frogs

97. White-lipped Frog—*Leptodactylus fragilis*
- Two males alternating calls; one very close. May 14, 1996, 9:17 pm, Panama Canal Zone, Panama. Erik D. Lindquist (Borror Laboratory of Bioacoustics).

Tailed Frogs—Ascaphidae

Ascaphus—Tailed Frogs

98a. Rocky Mountain Tailed Frog—*Ascaphus montanus*
- No recording; this species does not vocalize.

98b. Pacific Tailed Frog—*Ascaphus truei*
- No recording; this species does not vocalize.

Burrowing Toads—Rhinophrynidae

Rhinophrynus—Burrowing Toads

99a. Mexican Burrowing Toad—*Rhinophrynus dorsalis*
- Calls of one male. Recorded in Mexico by W. Frank Blair (Texas Natural Science Center).
- Large chorus. July 25, 2008, 11:20 pm, El Sauz, Texas. Michael L. Treglia.

Tongueless Frogs—Pipidae

Xenopus—Clawed Frogs

99b. African Clawed Frog—*Xenopus laevis*
- One male calling underwater. January 25, 1994. Recorded at the laboratory of John Gerhart, University of California, Berkeley. Carlos Davidson.

Cope's Gray Treefrog

Sources and Further Reading

We consulted many sources in the creation of this work and cannot include them all here. Below are listed some of our favorites, including field guides, reference works, and relevant organizations and their websites.

Field Guides

The following recently published field guides cover broad geographic areas. Not listed are a number of excellent guides to the frogs and toads of various states.

Robert C. Stebbins. 2003. *A Field Guide to Western Reptiles and Amphibians* (3rd ed.). Boston and New York: Houghton Mifflin Company. 533 pages.

Roger Conant and Joseph T. Collins. 1998. *A Field Guide to Reptiles and Amphibians of Eastern and Central North America* (3rd ed.). Boston and New York: Houghton Mifflin Company. 616 pages.

R. D. Bartlett and Patricia P. Bartlett. 2006. *Guide and Reference to the Amphibians of Eastern and Central North America (North of Mexico).* Gainesville: University Press of Florida. 284 pages.

———. In press. *Guide and Reference to the Amphibians of Western North America (North of Mexico) and Hawaii.* Gainesville: University Press of Florida.

Lawrence L. C. Jones, William P. Leonard, and Deanna H. Olson (eds.). 2005. *Amphibians of the Pacific Northwest.* Seattle, WA: Seattle Audubon Society. 227 pages.

Reference Works

We recommend the following technical references for those who are serious about the study of frogs and toads.

Kentwood D. Wells. 2007. *The Ecology and Behavior of Amphibians.* Chicago and London: University of Chicago Press. 1,148 pages.

Michael Lannoo. 2005. *Amphibian Declines: The Conservation Status of United States Species.* Berkeley: University of California Press. 115 pages.

F. H. Pough, Robin M. Andrews, John E. Cadle, Martha L. Crump, Alan H. Savitsky, and Kentwood E. Wells. 2003. *Herpetology* (3rd ed.). Essex: Benjamin Cummings. 736 pages.

H. Carl Gerhardt and Franz Huber. 2002. *Acoustic Communication in Insects and Anurans: Common Problems and Diverse Solutions.* Chicago and London: University of Chicago Press. 531 pages.

Keeping Frogs and Toads as Pets

R. D. Bartlett and Patricia Bartlett. 2007. *Frogs, Toads, and Treefrogs: A Complete Pet Owner's Manual.* Hauppauge, NY: Barron's Educational Series, Inc. 96 pages.

Nonprofit and Governmental Websites

The Center for North American Herpetology (CNAH): www.cnah.org

Society for the Study of Amphibians and Reptiles (SSAR): www.ssarherps.org

National Amphibian Atlas: www.pwrc.usgs.gov/naa

AmphibiaWeb: www.amphibiaweb.org

Partners in Amphibian and Reptile Conservation (PARC): www.parcplace.org

Amphibian Conservation Action Plan: www.amphibians.org

Volunteer Monitoring Programs

North American Amphibian Monitoring Program (NAAMP): www.pwrc.usgs.gov/naamp

Frogwatch USA: www.frogwatch.org

Frogwatch Canada: www.frogwatch.ca

Regional Websites

Regional websites focusing on frogs and toads are proliferating. We advise that you check with your state's department of conservation to see if they sponsor a site. And be sure to perform an Internet search using "frogs and toads" and your state or region as keywords. Two fine websites that we found particularly useful with respect to western frogs are www.californiaherps.com and www.reptilesofaz.com.

Audio Guides

A number of audio guides to the calls of frogs and toads in various states are available, with new ones being published every year. Check with your state's conservation department to see if one is available. The compact disc that accompanies this book is in part a combination of the contents of the following three guides, significantly improved by the addition of many excellent new recordings by the authors and other recordists:

Lang Elliott. 2004. *The Calls of Frogs and Toads.* Mechanicsburg, PA: Stackpole Books. 65-page full-color book and compact disc. This guide is a revised and expanded edition of the original title published by NatureSound Studio in 1992.

Carlos Davidson. 1995. *Frog and Toad Calls of the Pacific Coast: Vanishing Voices.* Ithaca, NY: Cornell Laboratory of Ornithology. 27-page booklet and compact disc.

———. 1996. *Frog and Toad Calls of the Rocky Mountains: Vanishing Voices.* Ithaca, NY: Cornell Laboratory of Ornithology. 27-page booklet and compact disc.

Spring Peeper

Index to Species and Groups

This index refers to page numbers and includes both common and scientific names, along with major groups. Bolded page numbers indicate species profiles. Refer to the Species and Track List on pages 40–41 for a list of species arranged by taxonomic groups and keyed to species numbers that are coordinated with the compact disc. The Contents on page 7 should be used as a general subject index.

Barking Treefrog

Relax to the Choruses of the Frogs

In late winter, spring, and early summer, warm and humid weather gives rise to an explosion of activity in the amphibian world, as frogs and toads appear like magic to breed in flooded pools, ponds, and marshes. Their choruses resound from wetlands across the countryside. Featuring a variety of superb stereo recordings from throughout the southeastern states, this compact disc conveys the natural mix of sounds. There is no narration—only the pure, unadulterated calls of our native frogs and toads. This disc is the perfect complement to the identification CD that accompanies this book. To learn more, visit www.frogconcertos.com.

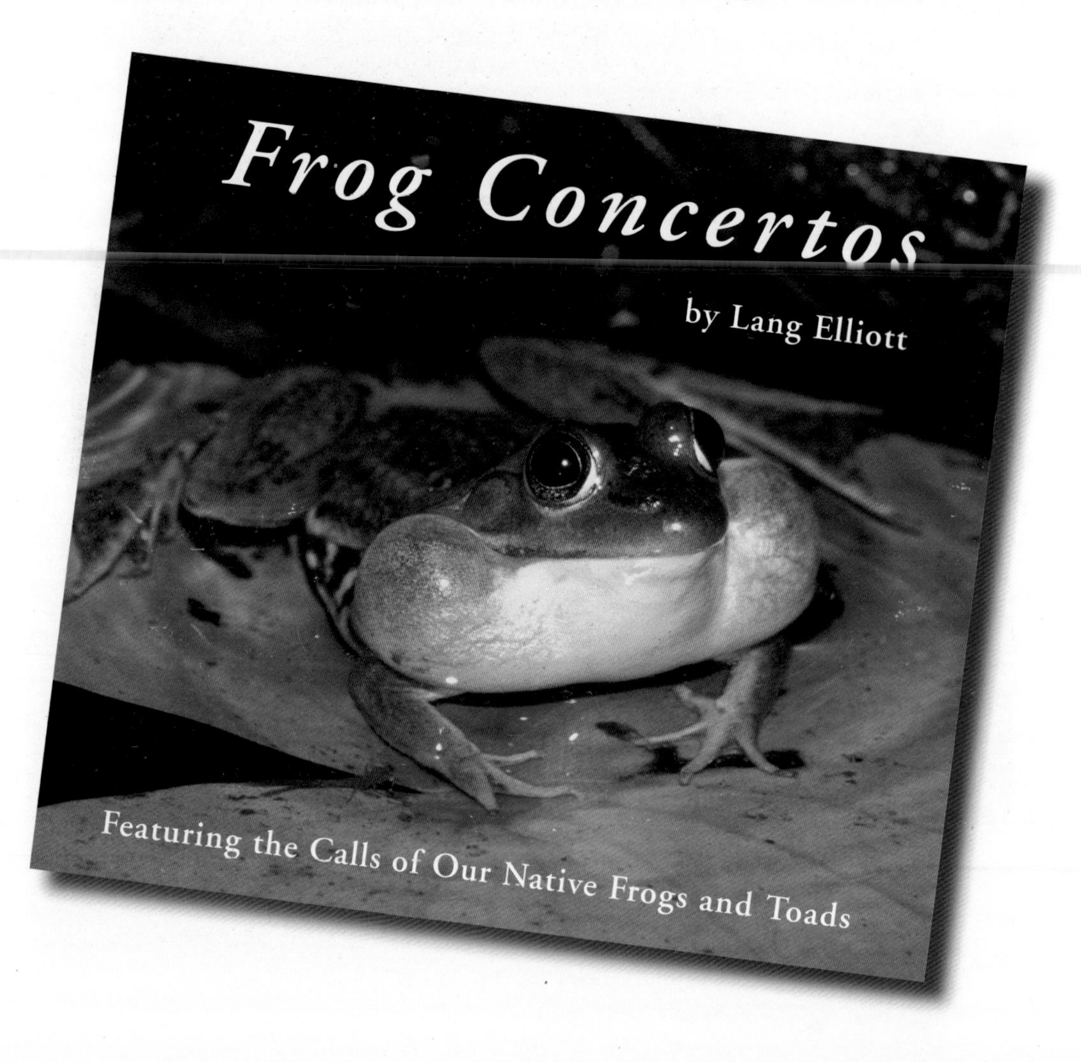